# Experimental and Evidence Based Protocols in Pharmacology

NIPA® GENX ELECTRONIC RESOURCES & SOLUTIONS P. LTD.
New Delhi-110 034

# About the Editors

**Prof. (Dr.) A.K. Srivastava** did his graduation and post-graduation in Veterinary Pharmacology from College of Veterinary Science, Mathura. He obtained his doctoral degree in Veterinary Pharmacology and Toxicology from Punjab Agricultural University (PAU), Ludhiana and Post Doctorate Fellowship from Institute of Toxicology, Munchen, Germany. He has also been honoured with Degrees of Doctor of Science (D.Sc. Honoris Causa) from various universities. He served as Chairman, Agricultural Scientists Recruitment Board (ASRB), DARE, Ministry of Agriculture, Government of India, New Delhi, Director cum Vice Chancellor, ICAR-National Dairy Research Institute, Karnal; Dean Faculty of Veterinary Science and Animal Husbandry and Director Resident Instructions and Dean Post graduate Studies at Sher-e-Kashmir University of Agricultural Sciences and Technology (SKUAST-J), Jammu; Head, Department of Veterinary Pharmacology and Toxicology and Controller of Examinations at PAU, Ludhiana. He Prof. Srivastava is a distinguished Member of National Academy of Sciences. He is feathered with fellow of National Academy of Agricultural Sciences, National Academy of Dairy Sciences and National Academy of Veterinary Sciences. He has been decorated with numerous prestigious awards and honors viz. ICAR Jawaharlal Nehru Award, International NOCIL Award, National Alarsin Awards in 1987-88 and in 1999-2000, ASET Gold Medal, German Academic Exchange Award, Gold Medals of Indian Science Congress Association and National Education Award 2016 etc. He is recipient of Dr. B.D. Garg outstanding Pharmacologist Award, Dr. V. Kurien Memorial Award, Chellappa Memorial Oration Award, Mansinhbhai Memorial Oration Award etc. He was also associated with working group of Planning Commission on "Animal Husbandry and Dairying" for XII Five Year Plan and Mid Term Evaluation of XI Five Year Plan, Government of India as Chairman. He led the DST-India delegation to New Zealand. He also represented India and was Chairman/ Member of Indian scientific delegation to Kansas University, USA, Belgium, Dubai, Ethiopia, Sri Lanka, Hong Kong, Nairobi, UK, Australia, New Zealand, Bangkok, Michigan State University, USA etc. He has more than 250 research papers, 10 books and 12 "National Policy Papers" in his credit and has guided more than 25 Ph.D. and post graduate students.

**Prof. (Dr.) Atul Prakash** did his B.V.Sc.& A.H. from Harbinger of Green Revolution G.B. Pant University of Agriculture and Technology, Pantnagar, Uttarakhand in the year 2004 and M.V.Sc. in Veterinary Pharmacology and Toxicology from N.D University of Agriculture and Technology, Ayodhya in the year 2006. Further he conferred with Doctor of Philosophy in Veterinary Pharmacology from Indian Veterinary Research Institute, Izatnagar, Bareilly in the year 2009. He has more than 15 years of experience in teaching of Pharmacology and Toxicology. He is awarded with several awards *viz.* Foreign Travel Grant from DBT, Govt. of India and CSIR, Govt. of India etc. He is also awarded with certificate of recognition from various scientific societies. He has guided more than 10 postgraduate research scholars. Currently, he is serving as Professor and Head in the Department of Veterinary Pharmacology and Toxicology, C.V.Sc. & A.H., U.P. Veterinary University (DUVASU) Mathura.

**Dr. Amit Shukla**, did his B.V.Sc. & A.H. and M.V.Sc. in Veterinary Pharmacology and Toxicology from G.B. Pant University of Agriculture and Technology, Pantnagar, Uttarakhand in the year 2010 and 2012, respectively. He completed his Ph.D. in Veterinary Pharmacology from Indian Veterinary Research Institute (IVRI), Izatnagar, Bareilly in the year 2015. He has more than 7 years of teaching experience in the field of Veterinary Pharmacology and Toxicology. He was awarded with several award *viz.* Best Educationist Award, CARS Project Award from DRDO-INMAS, Ministry of Defence, Government of India, New Delhi, Best Poster Award and Best presentation Award etc. Presently, he is working as an Assistant Professor in the Department of Veterinary Pharmacology and Toxicology, College of Veterinary Science and Animal Husbandry, UP Pandit Deen Dayal Upadhyaya Pashu Chikitsa Vigyan Vishwavidyalaya Evam Go Anusandhan Sansthan (DUVASU), Mathura- 281001, Uttar Pradesh, India.

**Dr. Soumen Choudhury** did his B.V.Sc. & A.H. from CAU, Imphal in the year 2005 and M.V.Sc. in Veterinary Pharmacology and Toxicology from U.P. Veterinary University (DUVASU), Mathura in the year 2007. He completed his Ph.D. in Veterinary Pharmacology from Indian Veterinary Research Institute (IVRI), Izatnagar, Bareilly in the year 2014. He has more than 12 years of teaching experience in the field of Veterinary Pharmacology and Toxicology. He was awarded with several award *viz.* Prof. V.V. Ranade Young Scientist Award by Indian Society of Veterinary Pharmacology & Toxicology (ISVPT) etc. Presently, he is working as an Associate Professor in the Department of Veterinary Pharmacology and Toxicology, College of Veterinary Science and Animal Husbandry, UP Pandit Deen Dayal Upadhyaya Pashu Chikitsa Vigyan Vishwavidyalaya Evam Go Anusandhan Sansthan (DUVASU), Mathura-281001, Uttar Pradesh, India.

# Experimental and Evidence Based Protocols in Pharmacology

The Protocol Series: Volume 05

**A. K. Srivastava**
**Atul Prakash**
**Amit Shukla**
**Soumen Choudhury**

**NIPA® GENX ELECTRONIC RESOURCES & SOLUTIONS P. LTD.**
New Delhi-110 034

**NIPA® GENX ELECTRONIC**
**RESOURCES & SOLUTIONS P. LTD.**
101,103, Vikas Surya Plaza, CU Block
L.S.C. Market, Pitam Pura, New Delhi-110 034
Ph : +91-11-43860225, Mob.: +91 9717133558, 9540816132
E-mail: newindiapublishingagency@gmail.com
Website: www.nipaersources.com



Print ISBN: 978-93-58874-40-2
ebook ISBN: 978-93-58878-36-3



Composed and Designed by NIPA®.

# Preface

**"Experimental and Evidence based Protocols in Pharmacology"** provides a holistic coverage of scientific protocols involved in the pharmacology and toxicological studies. The book serves as a reference to the biomedical researchers working in pharmacology and allied subjects to mitigate the challenges in the routine laboratory procedures. Book has 22 chapters with evidence based protocols and procedures to conduct the defined experiment. Further, this book provides single reference window with evidence based practical approaches to cater the need of academician and research scholars. The book is designed in a diligent manner to provide majority of protocols viz. Primary cell culture, Mesenchymal stem cells isolation and its application, DNA fragmentation Assay, oxidative stress evaluation protocols, Cytotoxicity and Genotoxicity Assays, Apoptosis studies, Analytical studies involving HPLC, GC-MS, ICP-OES and experimental pharmacological tool in smooth muscle pharmacological studies etc. to cement the scientific acumen. This book imbibes all the basic and advanced research protocols along with their translational approaches in experiments involved in pharmacology and also provides the insight of example based drug assays to help in bioprospecting of naive molecules.

Further, this book also provide a reference book to the researchers involved in toxicological testing using OECD testing guideline to calculate the lethal dose of any compound in a better manner. This MS also suffice the different methods to calculate the median lethal dose ($LD_{50}$). The important reproductive toxicity assays to ascertain the toxic potential of any xenobiotics is also incorporated in this MS. Moreover, sophisticated technique like Flow cytometry and other experimental models viz. Direct and Indirect LPS induced acute lung injury model has also well written to provide an insight of the experimentation involving laboratory animals.

**A. K. Srivastava**
**Atul Prakash**
**Amit Shukla**
**Soumen Choudhury**

# Contents

# List of Contributors

**A.K. Srivastava,** Vice Chancellor, UP Pandit Deen Dayal Upadhyaya Pashu Chikitsa Vigyan Vishwavidyalaya Evam Go Anusandhan Sansthan (DUVASU), Mathura- 281001, Uttar Pradesh, India.

**Afroz Jahan,** Assistant Professor, Department of Veterinary Pharmacology and Toxicology, College Of Veterinary Science, Rampura Phul, Guru Angad Dev Veterinary & Animal Sciences University, Rampura Phul Bathinda, Punjab – 151103.

**Amit Shukla,** Assistant Professor, Department of Veterinary Pharmacology and Toxicology, College of Veterinary Science and Animal Husbandry, UP Pandit Deen Dayal Upadhyaya Pashu Chikitsa Vigyan Vishwavidyalaya Evam Go Anusandhan Sansthan (DUVASU), Mathura- 281001, Uttar Pradesh, India.

**Anshuk Sharma,** Scientist, Division of Pharmacology and Toxicology, ICAR-Indian Veterinary Research Institute, Izatnagar, Bareilly 243122, Uttar Pradesh.

**Atul Prakash,** Professor and Head, Department of Veterinary Pharmacology and Toxicology, College of Veterinary Science and Animal Husbandry, UP Pandit Deen Dayal Upadhyaya Pashu Chikitsa Vigyan Vishwavidyalaya Evam Go Anusandhan Sansthan (DUVASU), Mathura- 281001, Uttar Pradesh.

**G.S. Rao,** Department of Veterinary Pharmacology and Toxicology, NTR College of Veterinary Science, Gannavaram Andhra Pradesh.

**Manjinder Sharma,** Department of Veterinary Physiology and Biochemistry, COVS, Ludhiana, Guru Angad Dev Veterinary and Animal Sciences University, Ludhiana, Punjab

**Meemansha Sharma,** Scientist, Division of Pharmacology and Toxicology ICAR-Indian Veterinary Research Institute, Izatnagar, Bareilly 243122, Uttar Pradesh.

**Milindmitra Lonare,** Associate Professor, Department of Veterinary Pharmacology & Toxicology, College of Veterinary Science, Guru Angad Dev Veterinary & Animal Sciences University, Rampura Phul Bathinda, Punjab – 151103.

**Mukul Anand,** Associate Professor, Department of Veterinary Physiology, College of Veterinary Science and Animal Husbandry, UP Pandit Deen Dayal Upadhyaya Pashu Chikitsa Vigyan Vishwavidyalaya Evam Go Anusandhan Sansthan (DUVASU), Mathura- 281001, Uttar Pradesh.

**Preeti Singh;** Assistant Professor, Department of Veterinary Pharmacology and Toxicology, College of Veterinary Science and Animal Husbandry, UP Pandit Deen Dayal Upadhyaya Pashu Chikitsa Vigyan Vishwavidyalaya Evam Go Anusandhan Sansthan (DUVASU), Mathura- 281001, Uttar Pradesh.

**Rajesh Mandil,** Professor and Head, Department of Veterinary Pharmacology and Toxicology, College of Veterinary & Animal Science, Sardar Vallabhbhai Patel University of Agriculture & Technology (SVPUAT), Meerut-250 110, Uttar Pradesh.

**S.P. Singh,** Professor, Department of Veterinary Pharmacology and Toxicology, College of Veterinary and Animal Sciences, GBPUAT, Pantnagar, India.

**S.P. Singh,** Associate Professor, Department of Animal Genetics and Breeding, College of Veterinary Science and Animal Husbandry, UP Pandit Deen Dayal Upadhyaya Pashu Chikitsa Vigyan Vishwavidyalaya Evam Go Anusandhan Sansthan (DUVASU), Mathura-281001, Uttar Pradesh.

**Sakshi Tiwari,** Assistant Professor, Department of Veterinary Pathology, College of Veterinary Science and Animal Husbandry, UP Pandit Deen Dayal Upadhyaya Pashu Chikitsa Vigyan Vishwavidyalaya Evam Go Anusandhan Sansthan (DUVASU), Mathura-281001, Uttar Pradesh.

**Sanweer Khatoon,** Department of Veterinary Parasitology, College of Veterinary and Animal Sciences, Navania, RAJUVAS, Bikaner, Rajasthan.

**Shalini Vaswani,** Assistant Professor, Department of Animal Nutrition, College of Veterinary Science and Animal Husbandry, UP Pandit Deen Dayal Upadhyaya Pashu Chikitsa Vigyan Vishwavidyalaya Evam Go Anusandhan Sansthan (DUVASU), Mathura-281001, Uttar Pradesh.

**Sivaraman Ramanarayanan,** Department of Veterinary Pharmacology and Toxicology, COVS, Kwashanganj, BASU, Patna, Bihar.

**Soumen Choudhury,** Assistant Professor, Department of Veterinary Pharmacology and Toxicology, College of Veterinary Science and Animal Husbandry, UP Pandit Deen Dayal Upadhyaya Pashu Chikitsa Vigyan Vishwavidyalaya Evam Go Anusandhan Sansthan (DUVASU), Mathura- 281001, Uttar Pradesh, India.

**Thakur Uttam Singh,** Principal Scientist, Division of Pharmacology and Toxicology, ICAR-Indian Veterinary Research Institute, Izatnagar, Bareilly 243122 Uttar Pradesh.

# 1

# Preparation of Primary Cell Cultures from Different Tissues/Organs

*Atul Prakash, Rajesh Mandil, Amit Shukla and A.K. Srivastava*

Cell culture refers to removal of cells from an animal or plant and their subsequent growth in a favorable artificial environment. The cells may be removed from the tissue directly and disaggregated by enzymatic or mechanical means before cultivation, or they may be derived from a cell line or cell strain that has already been establwashed.

Primary culture refers to the stage of culture after the cells are isolated from tissue and proliferated under appropriate conditions until they occupy all of the available substrate (i.e., reach confluence). At this stage, the cells have to be subcultured (i.e passaged) by transferring them to a new vessel with fresh growth medium to provide more room for continued growth.

Pharmaceutical and chemical industries depend largely on toxicological evaluations to provide information regarding safety of their commercial products. Toxicologists so far have been relying on animal studies to investigate potentially beneficial or harmful effects of drugs and chemicals prior to testing them in humans. Traditionally, the safety assessment process during drug development has been based on screening indiscriminately a plethora of new chemical entities (NCEs) using large number of animals as testing models in order to identify new therapeutic agents with the expectation that one or a few of these NCEs have some competitive advantages over the existing therapeutic agents. That is why they are more efficacious, bioavailable, selective, and safer. This approach is costly, time-consuming, and requires large quantities of test material. Currently, it takes about 10-12 years and a huge expenditure for a drug to move from research laboratory to patent. Thus alternative cell culture models in the early phases of drug discovery process can save time and money both by eliminating least beneficial NCEs.

## A. Isolation of Peripheral Blood Lymphocytes

### Requirements

- Heparin 100 IU/ml in PBS
- Sterile PBS (pH 7.2)
- Histopaque-1077
- RPMI-1640 Media

### Protocol

1. Collect 5 ml of blood from the test animal in 1 ml (20 IU/ml of blood) of diluted heparin.
2. Add 2 ml of PBS.
3. Layer over the diluted blood on 5 ml of histopaque-1077 in 15 ml centrifuge tube.
4. Centrifuge at 1500 rpm for 40 min.
5. Collect the buffy coat at the interface of plasma and histopaque in separate centrifuge tubes.
6. Wash the cells thrice with PBS and suspend in 2 ml of RPMI-1640 media supplemented with 10% foetal calf serum.
7. Determine the percentage cell viability by 0.1% trypan blue exclusion test using hemocytometer.
8. Finally dilute the cell suspension to $10^6$ cells/ml in RPMI-1640 medium.

## B. Isolation of murine thymocytes

### Requirements

- PBS
- Histopaque 1077
- RPMI-1640 Medium
- RBC lysis Buffer

### Protocol

1. Sacrifice healthy mice by cervical dislocation and aseptically remove the thymus.
2. Disintegrate the thymus quickly into many pieces.
3. With the plunger of a 10 ml glass syringe, the minced tissue is forced through the mesh into a petridish containing chilled PBS.
4. Once the entire tissue passes through, the cell suspensions are pipetted 4-5 times to break large cell clumps.

5. Collect the suspension into 15 ml centrifuge tube and allow standing on ice for 5 min.
6. Collect the top 12 ml of suspension into another centrifuge.
7. Centrifuge at 1500 rpm for 10 min to obtain the cell pellet.
8. Resuspend the pellet in PBS and layer over histopague-1077 in a ratio of 1:1.
9. Centrifuge at 1600 rpm for 40 min and collect the interface ring rich in thymocytes.
10. The collected interface is treated with 5 ml of RBC lysis Buffer (4.15g $NH_4Cl$; 0.5g $NaHCO_3$; 0.0186g $Na_2$-EDTA; 200ml DW) for 10 min.
11. Give two washings with PBS at 1200 rpm for 10 min.
12. Dissolve the pellet in 5 ml of RPMI -1640 medium with 10% FCS (growth medium; GM).
13. Viability count is done by 0.1% trypan blue dye and adjusts the cell density to get the $10^6$ cells/ml in RPMI-1640-GM.

## C. Isolation of Splenocytes

### Requirements

- PBS (pH 7.2)
- Histopaque-1077
- RPMI-1640 Media

### Protocol

1. Aseptically separate the spleen from the animal in ice cold PBS.
2. Place the spleen on a sterile, autoclaved nylon mash prewetted with PBS.
3. Remove the capsule of spleen using a pair of forceps and needles.
4. Disintegrate the spleen into many pieces with the plunger of a 10 ml glass syringe.
5. Force the tissue through the mesh into a petridish containing chilled PBS.
6. Pipette the cell suspensions 4-5 times to break large cell clumps.
7. Collect the cell suspensions in 15 ml centrifuge tubes and allow to stand on ice for 5 min.
8. Collect the top 12 ml of the suspensions into another set of centrifuge tube and pellet the cells by centrifugation (1500 rpm for 10 min).

9. Resuspend the cell pellet in PBS and layer over histopague-1077 in the ratio of 1:1.
10. Centrifuge the tubes at 1600 rpm for 40 min and collect the interface ring rich in splenocytes.
11. Wash the cells with PBS thrice and adjust the cell concentration to $10^6$ cells/ml in RPMI-growth medium after assessing the cell count and cell viability using trypan blue.

## D. Isolation and preparation of Hepatocyte cell culture

### Culture media

For culture of hepatocytes, medium M199 (GibcoBRL, USA) is used. One pouch of the dry media is dissolved in 1 liter of filtered distilled water. pH of the media is adjusted around 7.4 by adding 0.4% sodium bicarbonate. The media is then sterilized by filtration through seitz filter (0.22 mm) and stored in 50 ml sterilized conical flasks.

### Hanks Balanced Salt Solution (HBSS)

For preparation of HBSS, one vial of dry powder is dissolved and properly mixed in 1000 ml of sterilized distilled water. Sterilization is done through filtration. The prepared HBSS may be kept in 100 ml sterilized conical flask for later use.

### Other Accessories

For preparing of tissue culture, tissue culture flask (25 ml), tissue culture plates of 6, 24 and 96 well may be used depending on the parameter to be studied. All the glassware to be used should be properly washed and sterilized.

### Protocol

For isolation of mammalian hepatocytes, new born animals, preferably newly born rat pups are used. Pups are sacrificed by decapitation and the visceral cavity is openend immediately with a pair of sharp scissors aseptically. Viscera is shifted to one end and liver carefully removed from the body and put into a petridish having HBSS. After removing the undesirable tissue pieces and gall bladder, liver is transferred and washed in petridish with HBSS. After proper washing in HBSS, the liver piece is transferred to a petridish containing sufficient volume of media (M-199). Similar procedure is followed for all the animals sacrificed. The number of animals used depends on the amount of final cell suspension required.

## Trypsinization

Liver pieces are then passed through the barrel of a 5 ml sterilized fresh syringe to get it minced finely. The minced liver tissue is now subjected to trypsinization. Trypsin solution (1:250) is used for disintegration of the tissue. About 5 ml of trypsin per tissue piece is sufficient for one step of trypsinization. Trypsinization is carried out two times on the same tissue by changing the trypsin solution. The mixture of tissue in trypsin solution and HBSS is put on magnetic stirrer till tissue debris disintegrates almost completely. The suspension so obtained is then filtered through a muslin cloth tied over a glass funnel.

## Isolation of Hepatocytes

The cellular filtrate obtained after filtration of trypsinized tissue suspension is centrifuged to wash and obtain a pellet of hepatocytes. For this, the filtrate is washed twice with HBSS at 3000 rpm for 10 minutes in refrigerated centrifuge, third and final washing is done in media M-199. Pellet obtained after centrifugation, is mixed in 1 ml of media and the volume of the cell pellet obtained is measured. The final cell suspension is prepared by diluting the pellet 200 times with media. The suspension is observed under the microscope after staining with trypan blue dye to determine the live cell percentage.

## Seeding Rate

Culture plates are then seeded with the required volume of cell suspension depending on the type of plate used. The concentration of cells in the cell suspension is kept so as to obtain 1 X $10^4$ cells per ml. The plate or bottle is then incubated at $37^0$C with 5 % $CO_2$ pressure in a $CO_2$ incubator for the formation of monolayer of hepatocytes.

## Suggested Readings

Berry, M.N. and Friend, D.S. 1969. High-yield preparation of isolated rat liver parenchymal cells. J.Cell.Biol. 43: 506-519.

Ekwall, B. 1980. Screening of toxic compounds in tissue culture. Toxicology. 17; 127-142.

Fuss IJ, Kanof ME, Smith PD, Zola H. 2009. Isolation of whole mononuclear cells from peripheral blood and cord blood. Curr Protoc Immunol. Apr;Chapter 7:Unit7.1. doi: 10.1002/0471142735.im0701s85

Kim, Y. M., Bergonia, H. A., Muller, C., Pitt, B. R., Watkins, W. D., Lancaster Jr., J. R. 1995. Loss and degradation of enzyme-bound heme induced by cellular nitric oxide synthesis. J. Biol. Chem. 270, 5710– 5713.

Nardone, R.M. 1980. Tissue culture systems in toxicity testing. In A.N.Rowan and C.J.Stratman (Eds), The use of Alternatives in Drug Research, The Macmillin Press, London.

Stammatu, A.P., Silano, V and Zucco, F. 1981. Toxicology investigations with cell culture systems. Toxicology. 20 : 91-153.

# 2

# Procedures to Evaluate Cytotoxicity in Primary Cell Cultures

*Atul Prakash, Rajesh Mandil, Amit Shukla and A.K. Srivastava*

Environmental agents (xenobiotics) that are injurious to cells trigger a spectrum of responses that can range from cell death to adaptation, repair, and proliferation. Cell death, in turn, can take the form of apoptosis (a programmed form of cell death). Xenobiotics have been implicated in toxic cell injury, among them free radicals (nitrogen, oxygen, and carbon derivatives) are known to interact with cellular constituents including membrane lipids and chromagen materials.

## Cytotoxicity

**Cytotoxic:** Toxic to cells, cell-toxic, cell-killing. Any agent or process that kills cells. The prefix cyto- denotes a cell. It comes from the Greek "kytos" meaning hollow, as a cell or container. Toxic is from the Greek "toxikon" = arrow poison.

In nutshell, cell cytotoxicity refers to the ability of certain chemicals or mediator cells to destroy living cells. By using a cytotoxic compound, healthy living cells can either be induced to undergo necrosis (accidental cell death) or apoptosis (programmed cell death).

Measuring cytotoxicity can prove to be very valuable tool in identifying compounds that might pose certain health risks in humans and animals. Cytotoxicity assays are widely used by pharmaceutical industry to screen for cytotoxicity in compound libraries from initial high-throughput drug screens for unwanted cytotoxicity. This can be of vital importance and prove to be quite indispensable during research phase of developing therapeutic new pharmaceutical products to ensure the safety of end-users. Additionally, understanding the mechanisms involved in cytotoxicity can likewise give researchers a more in-depth knowledge on the biological processes (both normal and abnormal) governing cell growth, cell proliferation and death. Compounds that have cytotoxic effects often compromise cell membrane integrity. Therefore, assessing cell membrane integrity is one of the most common ways to measure cell viability and cytotoxic effects. There are three commonly used methods which can be employed to assess *in vitro* cytotoxicity (disruption of cell integrity) in cell cultures:

1. **Trypan blue dye uptake assay**
2. **Propidium iodide assay**
3. **Neutral red dye retention assay**

## 1. Trypan blue cell viability assay

This common cell viability assay is based on the ability of a cell with an intact membrane to exclude the dye trypan blue. Therefore, this assay allows us to differentiate between cells with intact and disrupted membranes.

### Requirements

- 0.1% Trypan Blue Dye (in PBS)
- Hemocytometer with Cover slips

### Protocol

1. Take 20 ml of diluted cell suspension in a PCR tube
2. Add 20 ml of 0.1% trypan blue dye
3. Mix well and immediately charge the hemocytometer and allow standing for 2 min.
4. See under 10x and count the live (bright) and dead (blue) cells in 64 sec. Square (4 large corner square)

### Calculate

Total viable cell/ml of culture = No of cells in 4 large square x $10^3$ x 2 (dilution factor)

% Viability = [Average no of viable cell x $10^4$ x DF]/ Total count (Viable + Non-Viable) x 100

% Non-Viability = [Average no of Non-Viable cell x $10^4$ x DF]/ Total count (Viable + Non-Viable) x 100 or (100 - % Viability)

## 2. Propidium iodide assay

The fluorescent dye propidium iodide (excitation 530 nm, emission 645 nm) is excluded from living cells, but enters and combines with double-stranded nucleic acids (DNA) of dead cells. The assay for cell viability using propidium iodide attachment to double-stranded nucleic acids can be done simultaneously with determination of membrane integrity using a variety of membrane-permeate fluorescent probes (e.g., CFDA-AM, which has an excitation 485 nm and emission 530 nm) because the fluorescent spectra do not overlap.

## Protocol

### *Materials*

a) $0.5 \times 10^5$ cells/ml
b) Test compounds: toxicant, vehicle and positive control (digitonin solution)
c) Phosphate buffer saline
d) 376 µM Digitonin solution
e) 30 µM propidium iodide dye
f) Flat bottomed, 96-well microtitre plates, for cell culture
g) Fluorescence microtitre plate reader

1. Add 10 µl of 376 µM Digitonin solution to positive control wells.
2. Add 100 µl propidium iodide (30 µM) to all wells.
3. Incubate the plate at 37°C for 30 min.
4. Read in a microtiter plate fluorescence spectrophotometer set with excitation at 530 nm and emission at 645 nm. (*Alternatively, excite at 488 nm.*)
5. Compare the wells exposed to test compounds and wells exposed only to vehicle (negative control) to wells exposed to Digitonin (100% cell death; the positive control). (*The positive control will provide wells with the highest fluorescence.*)
6. Calculate per cent of viable cells as fluorescence in test wells less negative control divided by fluorescence in Digitonin-treated wells less negative control, with the quotient multiplied by 100.

## 3. Neutral red dye retention assay

The basis of neutral-red dye uptake assay is the ability of viable cells to incorporate and accumulate the supravital dye neutral red in lysosomes. Cellular damage by the action of xenobiotics results in decreased uptake and retention of neutral red.

## Protocol

1. Incubate 100 µl cells (1 to $2 \times 10^5$ cells/ml) with 0.1% (w/v) neutral red for 3 hr at 37°C in microtitre plates.
2. Wash with PBS by adding 200 µl PBS to each well and mix by pipetting. Centrifuge the plate for 10 min at 1000 rpm with a plate carrier.
3. Carefully remove the PBS. Repeat the PBS wash two to three times.
4. Add 1% glacial acetic acid/50% methanol to extract the dye and fix the cells.

5. Read absorbance at 540 to 550 nm or fluorescence at 645 nm with 488 nm excitation

## Suggested Readings

Babich, H. and Borenfreund, E. 1992. Neutral red assay for toxicology in vitro. In In Vitro Methods of Toxicology (R.R. Watson, ed.) pp. 237-251. CRC Press, Boca Raton, Fla.

Barile, F.A. 1994. Introduction to In Vitro Cytotoxicology Mechanisms and Methods. CRC Press, Boca Raton, Fla.

Krwashan, A. 1975. Rapid flow cytometric analysis of mammalian cell cycle by propidium iodine staining. J. Cell Biol. 66:188-193.

Mosman, T. 1983. Rapid colorimetric assay for cellular growth and survival: Application to proliferation and cytotoxicity assays. J. Immunol.Methods 65:55-63.

O'Hare, S. and Atterwill, C.K. 1995. Methods in Molecular Biology 43: In Vitro Toxicity Testing Procedures. Human Press, Totowa, N.J.

# 3

# Genotoxicity Assays for Safety Evaluation of Chemicals in Cell Culture

*Atul Prakash, Rajesh Mandil and A.K. Srivastava*

Safety is a prime concern for humans and animals. Without evaluating the safety of a chemical, no regulatory authority in the world allows the chemical to be used by humans. There are stipulated guidelines for toxicity evaluations. Among the toxicity assays genotoxicity assays are important and have occupied centre stage due to increase in understanding of the genome.

Genotoxicity describes the property of chemical agents to damage the genetic information within a cell to cause mutations, which may even lead to cancer. While genotoxicity is often confused with mutagenicity, all mutagens are genotoxic; however, not all genotoxic substances are mutagenic. Generally and depending on the agent, a cytotoxic agent may be genotoxic (causing DNA damage), mutagenic (causing gene mutation) or it may be harmful for cell organelles (e.g. cell membrane) or it may even be teratogenic. Therefore, every cytotoxic agent is not genotoxic.

## Why we Need to Evaluate Genotoxicity?

Toxicity can be at any level according to the nature of chemicals and type of exposure. Duration of toxicity can also be short lived or long lived within a generation or continue to next generation. Genotoxicity effects are able to continue in next generation(s), in which the exposed population may not suffer seriously but the next generation may suffer without being exposed to the same chemicals. Therefore, genotoxicity evaluations need serious attention.

Genotoxic effects can be assessed directly by measuring the interaction of agents with DNA or more indirectly through assessment of DNA repair or the production of gene mutations or chromosomes. DNA damage emanates from the inherent chemical instability of nucleic acids, from errors made by polymerase during DNA replication as well as from exposure to DNA damaging agents present in the environment or produced by certain endogenous processes. Individual sensitivity to particular chemical mutagens can be establwashed by studying the influence of genetic polymorphisms on genotoxic response of cells in *in vitro* experiments. The most useful assays in analyzing genotoxic response are cytogenetic tests because they can be applied to any mutagen known to produce chromosomal alterations.

### Types of Available Assay Systems

There are 16 assays in the OECD guidelines for evaluating genotoxicity of chemicals. International Council for Hormonisation (ICH) guidelines also suggests a battery of tests with minimum three assays. The assays having regulatory acceptance can be categorized into three segments according to the nature of the end point.

### Assays for Gene Mutation

As the name suggests, these assays are being used considering the mutations in the gene or DNA of the organism. There are about five assay systems under this category having regulatory acceptance. These are as follows:

1. Reverse mutation assays using *Salmonella* sp. and *E. coli*
2. Gene mutation assay in mammalian cell
3. Drosophila sex linked recessive lethal assay
4. Gene mutation assay of *Saccharomyces cerevisiae*
5. Mouse spot test

In reverse mutation assay, *Salmonella* spp. and *E. coli* are being used with a mutation in histidine (*Salmonella* spp.) or tryptophan (*E. coli*) gene modification. The *Salmonella* reverse mutation assay is the widely used assay system in the globe. Gene mutation assay in mammalian cells used generally are L7518Y of mouse or CHO and V-79 cells of Chinese hamster. With Thymidine kinase, HPRT and sodium/ potassium ATPase targets are being used. In Drosophila sex linked assay inheritance is being utilized as a marker involving 80% of the X chromosomal loci. Gene mutation assay in *Saccharomyces cerevisiae* haploid and diploid strains are being used to measure the mutations. Mouse spot test is the only mammalian genotoxicity assay system which uses developing embryos for exposure of genotoxicant and later the phenotypically coat color is used as a indicator of genotoxicity.

### Assays for Chromosomal Aberrations

Chromosome being the visible genetic material (under microscope) attracted the attention of toxicologists for evaluation of genotoxicity. Following are the accepted assay systems to assess chromosomal aberrations:

1. *In vitro* cytogenetic assay
2. *In vivo* cytogenetic assay
3. Micronucleus assay

4. Dominant lethal assay
5. Heritable translocation assay
6. Mammalian germ cell cytogenetic assay

In *in-vitro* cytogenetic assay, usually human peripheral lymphocytes or suitable cell lines are being used whereas *in vivo* cytogenetic assay mouse is the model of choice although hamster and rat can also be used. Mammalian germ cell cytogenetic assay evaluates chromosomes in the germ cells of mouse. Micronucleus assay uses the exfoliation of nuclear material during development of red blood cells as the marker. Dominant lethal assay considers the rate of resorption of implanted embryos in the rodent system and heritable translocation assay considers phenotypic expression.

### Assays for DNA Effects

Genotoxicants ultimately affect the genetic material or DNA which can be measured by several assay systems. The following are the most practiced assay systems having regulatory clearance:

1. DNA damage and repair : Unscheduled DNA synthesis *in vitro*
2. DNA damage and repair : Unscheduled DNA synthesis *in vivo*
3. Mitotic recombination in *Saccharomyces cerevisiae*
4. *In vitro* sister chromatid exchange

The unscheduled DNA synthesis test measures the DNA repair synthesis after excision and removal of a stretch of DNA containing the region of damage induced by chemical or physical agents utilizing mammalian cells in culture or rat hepatocytes provided with some exogenous mammalian metabolic activation (*in vitro*) or the whole liver of the rodents. Mitotic recombination in *Saccharomyces cerevisiae* is based on the principle that mitotic crossing over and mitotic gene conversion can be detected in these organisms hence making the perturbations in the genetic material recognizable as a result of toxicity. *In vitro* sister chromatid exchange assay involves the detection of reciprocal exchanges of DNA between two sister chromatid of a duplicating chromosome. Sister chromatid exchanges (SCEs) can be measured in mammalian and non-mammalian systems.

The evaluation of genotoxicity of the chemicals is necessary for getting a clearance from the regulatory authority. From the above described assays, no single assay is able to answer the queries of investigator about the nature of genotoxicant and the mechanism. Therefore, it is always advisable to conduct a battery of assays with relevant end-points so as to reach to a decisive conclusion.

## Further Readings

Babich, H. and Borenfreund, E. 1991. Cytotoxicity and genotoxicity assays with cultured fwash cells: A review. Toxicol. In Vitro 5:91-100.

Dunkel, V.C., Rogers, C., Swierenga, S.H.H., Brillinger, R.L., Gilman, J.P.W., and Nestman, E.R. 1991. Recommended protocols based on a survey of current practice in genotoxicity testing laboratories: III. Cell transformation in C3H/10T1/2 mouse embryo cell, BALB/c3T3 mouse fibroblast and Syrian hamster embryo cell cultures. Mutat. Res. 246:285-300.

Edler, L. 1994. Biostatistical issues in the design and analysis of multiple or repeated genotoxicity assays. Environ. Health Perspect. 102 (Suppl. 1):53-59.

Fenech, M. 1993. The cytokinesis-block micronucleus technique: A detailed description of the method and its application to genotoxicity studies in human populations. Mutat. Res. 285:35-44.

Kramers, P.G.N., Knaap, A.G.A.C., van der Heijden, C.A., Taalman, R.D.F.M., and Mohn, G.R. 1991. Role of genotoxicity assays in the regulation of chemicals in The Netherlands: Considerations and experiences. Mutagenesis 6:487-493.

Li, A.P., Aaron, C.S., Auletta, A.E., Dearfield, K.L., Riddle, J.C., Slesinski, R.S., and Stankowski, L.F., Jr. 1991. An evaluation of the roles of mammalian cell mutation assays in testing of chemical genotoxicity. Regul. Toxicol. Pharmacol.14:24-40.

Naji-Ali, F., Hasspieler, B.M., Haffner, D., and Adeli, K. 1994. Human bioassays to assess environmental genotoxicity: Development of a DNA repair assay in HepG2 cells. Clin. Biochem.27:441-448.

Swierenga, S.H.H., Bradlaw, J.A., Brillinger, R.L., Gilman, J.P.W., Nestmann, E.R., and San, R.C. 1991. Recommended protocols based on a survey of current practice in genotoxicity testing laboratories I. Unscheduled DNA synthesis assay in rat hepatocyte cultures. Mutat. Res.246:235-253.

# 4

# DNA Fragmentation Assay

***Rajesh Mandil, Atul Prakash, S.P. Singh and Preeti Singh***

Apoptosis is triggered by extrinsic and intrinsic stimuli as radiation, oxidative stress and genotoxic chemicals. It is defined by characteristic changes in the nucleus morphology including chromatin condensation and fragmentation, overall cell shrinkage/rounding and formation of apoptotic cell bodies. Among the many markers of apoptosis, DNA fragmentation is considered a hallmark. The DNA ladder assay described here is a simple, sensitive, cost-effective and rapid method for estimating apoptosis in cells.

**Preparations of Reagents**

1. **2.7 % EDTA solution (pH-8.0)**

   | | |
   |---|---|
   | EDTA disodium salt | 2.7 g |
   | Double distilled water up to | 100 ml |

   Adjust pH 8.0 using NaOH pellets

   Sterilize by autoclaving and store at room temperature

2. **0.5 M EDTA solution (pH-8.0)**

   | | |
   |---|---|
   | EDTA disodium salt | 186.1 g |
   | Double distilled water up to | 1000 ml |

   Adjust pH 8.0 using NaOH pellets (approx. 20 g NaOH). (EDTA dissolve only at pH -8.0)

   Sterilize by autoclaving and store at room temperature

   (EDTA-Chelates $Mg^{++}$ ions, protects from nucleases and makes the plasma membrane more fragile).

3. **5 M NaCl**

   | | |
   |---|---|
   | NaCl | 29.22 g |
   | Double distilled water up to | 500.0 ml |

   Autoclaved and store at room temperature

4. **DNA extraction buffer**

| | |
|---|---|
| 1 M Tris buffer (pH-8.0) | 5 ml |
| 5 M NaCl | 40 ml |
| 0.5 M EDTA solution | 2 ml |
| Double distilled water up to | 500 ml |

Sterilize by autoclaving in batches of 100 ml and store at room temperature.

5. **10 % Sodium dodeycyl sulfate (SDS or sodium laurel sulfate)**

| | |
|---|---|
| SDS | 10 g |
| Autoclaved double distilled water up to | 100 ml |

Adjust pH 7.2 using conc. HCL. Heat in water bath at 60 °C to dissolve and then store at RT.

6. **3.0 M sodium acetate**

| | |
|---|---|
| Sodium acetate anhydrous | 24.6 g |
| Double distilled water up to | 100 ml |

Adjust pH 5.5 using glacial acetic avid

Autoclaved in batches of 20 ml.

Function of 3.0M sodium acetate to precipitates DNA

7. **Equilibrated Tris saturated phenol containing 8-hydroxyquinolone**

Function of phenol: RNA with poly A tail is dissolved in alkaline phenol, inhibits RNase and weakly chelates metal ion.

8. **Phenol: chloroform : isoamyl alcohol preparation (25:24:1)**

| | |
|---|---|
| Tris saturated phenol | 25 ml |
| Chloroform isoamyl alcohol (24:1) | 25 ml |

Mix thoroughly and store in amber coloured bottle at 4° C

9. **Chloroform : isoamyl alcohol preparation (24:1)**

| | |
|---|---|
| Chloroform | 24 ml |
| Isoamyl alcohol | 1 ml |

Mix thoroughly and store in amber coloured bottle at 4° C

10. **Absolute ethanol**

11. **70 % Ethanol**

| | |
|---|---|
| Ethanol | 70 ml |
| Autoclaved double distilled water up to | 30 ml |

Mix thoroughly and store in amber coloured bottle at 4° C

**12. 1 M Tris (pH-8.0)**

| | |
|---|---|
| Tris HCL | 157.6 g |
| Double distilled water up to | 1000 ml |

Adjusted pH 8.0 with NaOH pellets. Autoclaved in 100 ml batches.

**13. 1X TE buffer- pH 8.0 for DNA**

| | |
|---|---|
| 1 M Tris | 250 ul |
| 0.5 M EDTA (pH 8.0) | 50ul |
| Double distilled water up to | 25 ml |

Autoclaved and stored at 4° C

**14. PBS (pH 7.4)**

| | |
|---|---|
| NaCl | 8 g |
| KCl | 0.2 g |
| $Na_2HPO_4$ | 1.44 g |
| $KH_2PO_4$ | 0.24 g |
| $H_2O$ | 800 ml |

Adjust pH 7.4 with HCl and make final volume of 1 l, autoclaved and stored at RT.

## Reagents for Agrose Gel Electrophoresis

**1. Ethidium bromide ( 10 mg/ml)**

| | |
|---|---|
| Ethidium bromide | 10 mg |
| Autoclaved double distilled water up to | 1 ml |

Wrap in aluminum foil and stored in dark place at RT.

**2. 6 X loading dye**

| | |
|---|---|
| Bromophenol blue | 0.25 % |
| Xylene cyalone | 0.25 % |
| Sucrose in water | 40 % |

Mix and store at 4° C

**3. 5 X TBE buffer**

| | |
|---|---|
| Tris base | 54 g |
| Boric acid | 27.5 g |
| 0.5 M EDTA (pH -8.0) | 20 ml |
| Autoclaved double distilled water up to | 1000 ml |

Autoclaved and stored at Room Temperature.

## Isolation of Genomic DNA

Genomic DNA from primary cells culture of PMNC and splenocytes of rats is isolated by using standard phenol-chloroform DNA isolation protocol of Sambrook and Russel (2001) as described below:

1. Preparation of primary cells culture of different cells, namely-PBMNCs, thymocytes and splenocytes form Wistar rats and viability checked by trypan blue dye exclusion test.
2. $10^6$ cells/ml of primary cell culture is used and seeded in 24 well culture plates containing along with treatment with xenobiotics at different μM concentration.
3. Cell culture plates are incubated in $CO_2$ incubator at 37°C with 5% $CO_2$ for 12 hrs.
4. The cells are collected in 1.5 ml of micro centrifuge tube and centrifuged at 3200 rpm for 10 min at 4°C; supernatant is discarded and pellet is washed with PBS (pH 7.2).
5. DNA extraction buffer (500 μl/tube) is added in tubes. Gently tap to disperse the cell pellet in extraction buffer and keep in water bath for 1.0 hr at 37 °C (it is to ensure that pellet is completely suspended so that cells are accessible to SDS and Proteinase K).
6. 10 % SDS is added (20 μl/ml) to cell suspension and the tubes are gently mixed by inverting the tube. The contents of tube become viscous which indicates lysis of cells. After that sample is handled gently from onward to avoid shearing of DNA.
7. Proteinase K (15 μl) of 20 mg Proteinase K/ml of buffer is added in tubes in two pulse i.e. half the requirement is added to tube in 1$^{st}$ pulse, mixed gently end to end and kept in water bath at 50°C. After 3-4 hrs, the second pulse of the remaining amount of Proteinase K is added and tubes are incubated at 50°C overnight.
8. Equal volume of equilibrated phenol (Tris saturated phenol pH> 7.8 ) is added in the tubes next day morning. Mix the contents of tube gently by inverting the tube for 15 min till a light coffee coloured uniform solution without any balls of phenol is formed and then centrifuge at 3400 rpm for 15 min.
9. The upper aqueous phase containing DNA is transferred into fresh 1.5 ml clean and autoclaved micro centrifuge tube by means of wide bore of 1.0 ml tip. Upper aqueous phase is very viscous and care should be taken during transfer of aqueous phase so that lower organic phase (containing

phenol, cell lysate, proteins etc) and white interface (containing proteins) is not disturbed.

10. Similar extraction is done (as in above step) once with equal volume of phenol: chloroform: isoamyl alcohol (25:24:1) and with chloroform: isoamyl alcohol (24:1).
11. To the final aqueous phase obtained, add double the volume of chilled (-20°C) ethanol/isopropanol (room temperature) DGF. Tubes are mixed gently by inversion and kept at room temperature to allow precipitation of DNA.
12. DNA along with 500 µl ethanol is transferred into fresh micro centrifuge tube by means of wide bore tips or centrifuge at 10000 rpm for 10 min.
13. The supernatant is discarded by gentle inversion where DNA pellet is intact, otherwise aspirated.
14. The DNA pellet is washed twice with 500 µl of 70% ethanol i.e. 500 µl of 70% ethanol is added and Eppendorf tube is centrifuge at 10000 rpm for 10 min at RT.
15. Finally 70% ethanol is discarded and DNA pellet is air dried by inverting tube on blotting paper so that last traces of ethanol are removed, however, the pellet should not be over dried or else dissolution becomes difficult.
16. Approximately 50 µl of TE buffer is added and kept in water bath at 60°C for 2 h to inactivate DNAase or other enzymes.
17. Micro centrifuge tube with DNA is stored at 4°C for a week so that DNA is dissolved and subsequently at -20°C indefinitely.

Concentration and purity of DNA is determined spectrophometrically at OD 280 and 280 by Nanodrop. The integrity of DNA is examined by agarose gel (1.0%) electrophoresis and visualized by gel under UV light in Gel Documentation system after staining with ethidium bromide.

## Suggested Readings

Sambrook, J. and Russel, D.W. 2001. Molecular cloning: A Laboratory Manual. Cold/Spring Harbor Laboratory Press, New York, A1-A12

# 5

# Single Cell Gel Electrophoresis (SCGE/Comet Assay) in Primary Cell Culture

*Atul Prakash and Rajesh Mandil*

The concept of single cell micro-gel electrophoresis (SCGE)/comet assay is first introduced by Ostling and Johanson (1984) as a method to measure DNA single-strand breaks under neutral conditions. Singh *et al.* (1988) modified this technique by using alkaline conditions (>13 pH) of electrophoresis, which enabled detection of not only frank strands breaks but also alkali labile sites, DNA cross- linking and incomplete excision repair sites. Now-a-days, this technique is extensively used to study the genotoxic effects of various chemicals. The advantages of SCGE technique include: (1) collection of data at the level of individual cell, allowing for more robust types of statistical analysis; (2) need for small number of cells per sample (<10,000); (3) its sensitivity for detecting DNA damage; and (4) that virtually any eukaryotic cell population is amenable to analysis. The idea is to combine DNA gel electrophoresis with fluorescence microscopy to visualize migration of DNA strands from individual agarose-embedded cells. If the negatively charged DNA contained breaks, DNA supercoils are relaxed and broken ends are able to migrate toward the anode during a brief electrophoresis. If the DNA is undamaged, the lack of free ends and large size of the fragments prevented migration. Determination of the relative amount of DNA that migrated provided a simple way to measure the number of DNA breaks in an individual cell.

Genotoxic effects of any xenobiotics in PMNC and splenocytes of Wistar rats can be studied according to the working protocol of Dhawan *et al.* (2009) as described below:

## Preparation of Reagents

1. **Stock solution of 200 mM EDTA (8.0 pH):** Stock solution of 200 m MEDTA is prepared by dissolving 3.72 g of EDAT in 50 ml autoclaved DW and pH is adjusted to 8.0 with 40% sodium hydroxide (NaOH) pellets.

2. **Stock solution of 10 M (10N) NaOH:** 100 g NaOH pellets are dissolved in 250 ml autoclaved DW and kept in air tight container.
3. **Stock lysing solution:** Ingredients for 250 ml

| | |
|---|---|
| NaCl | 36.52 g |
| EDTA disodium salt | 9.3 g |
| Trizma | 0.3 g |
| NaOH | 2 g |

Add ingredients to about 230 ml autoclaved DW and begin stirring the mixture. Adjust the pH to 10.0 using concentrated HCl or NaOH and to make final volume 250 ml with DW Stored at room temperature.

**For final working lysing solution**: Add 1ml Triton X-100 and 10 ml DMSO in 89 ml stock lysing solution then refrigerate for at least 30 minutes prior to slide addition. Working lysis solution should prepared fresh each time. The purpose of the DMSO in the lysing solution is to scavenge radicals generated by the iron released from hemoglobin when blood or animal tissues are used. It is not needed for other situations or where the slides are to be kept in lysing for a brief time only.

**Electrophoresis buffer for alkaline comet assay:** It is prepared by dissolving

| | |
|---|---|
| 10 N NaOH | 30 ml |
| 200 mM EDTA | 5 ml |
| DW | add upto 1000 ml |

pH of this solution should be > 13.0

4. **Neutralization Buffer (pH -7.5)**: It is prepared by dissolving

| | |
|---|---|
| Tris buffer (Sigma) | 3 g |
| DW | 60 ml |

pH adjust pH to 7.5 with concentrated (>10 M) HCl: stored at 4° C.

5. **Staining Solution:** 10 mg ethidium bromide is added to 1.0 ml autoclaved DW. Working staining solution is prepared form the stock by taking 10 µl and add 4990 µl autoclaved DW in amber coloured vial.

## Preparation of Base Slides

1. Slide is dipped in methanol and then burnt them over blue flame to remove the grease, dust and oil.
2. Prepare 1.5% normal melting agarose (Sigma) (NMA-1.5 g per 100ml PBS) and 0.5% LMPA (25 mg per 5.0 ml PBS) in microwave or heat

until near boiling and the agarose dissolves. Place LMPA vial in a 40ºC water bath to cool and stabilize the temperature. While NMA agarose is kept at 100°C.

3. Cell isolation: Single cells of thymocytes and splenocytes from PMNC and spleen are isolated after cervical dislocation of rats of 120-150 g bw and viability checked by trypan blue dye exclusion test. The $5X10^6$ cells/well are kept for culturing and treated with xenobiotics and antioxidants, respectively, and kept for incubation of 12 hr in $CO_2$ incubator. The cells are collected in 1.5 ml of micro centrifuge and centrifuge at 3200 rpm for 10 min at 4°C, supernatant is discarded and pellet is washed with PBS (pH 7.2).
4. Preparation of first layer of agarose on slides by dipping conventional pre-cleaned slides for few seconds in 100 ml wide mouth beaker containing 1.5 % NMA up to one-third the area and gently removes. Wipe underside of slide to remove excess agarose and lay the slide in a tray on a flat surface to air dry. We generally prepare slides the day before use. The slides may be store at room temperature until needed; avoid high humidity conditions.
5. Preparation of second layer of agarose: Cell pellet of PMNC and splenocytes are uniformly mixed with 100 μl of 0.5% LMPA and then poured on first layer of agarose carefully at 3-4 drops with micropipette an immediately covering the drops with full length cover slip to avoid formation of air bubbles. Slides are properly coded with glass slide marker according to the treatment. These slides are kept on ice-packs for 15-20 min to solidify $2^{nd}$ layer of agarose.
6. After solidification, cover slips are removed and slides are kept in coupling jar containing freshly prepared lysis solution at 4°C in refrigerator for overnight.
7. Next day, slides are removed from lysis solution and kept for 30 min in freshly prepared electrophoretic buffer so as to cause unwinding of DNA and expression of alkali-labile sites.
8. Slide is run in horizontal electrophoresis (Bio Rad) chamber with same electrophoresis buffer (pH>13) at 25 V and 300 mA for 1.0 hrs for 1 h.
9. Neutralization: After running in electrophoresis chamber, slides are gently removed with the help of forceps and placed horizontally in a tray. Each slide is then covered with neutralizing buffer for 5 min and then decant it; same step is repeated three times to remove alkali and detergent. This step is critical to bring down pH from 13 to 7.5.

10. After neutralization, slides are stained by placing 3-4 drops of 100 μl working ethidium bromide solution at equal distance and immediately covered with cover slip.
11. Slides analysis: Analyze cells on fluorescent microscope, individual cell/ comets will be observed and images are captured at 40X magnification using green filter. Duplicate slides per treatment are made. At least 50 cells from each slide are scored and total of 100 cells per/treatment should be scored to get reproducible data.

The comet assay is extensively used in *in vitro* and *in vivo* genetic toxicology. Different types of cell lines and practically any type of cell from any target organs can be tested by comet assay.

## Cell Cycle Analysis

The comet assay can be used to study the DNA damage in cells in different cell cycle phases. The total comet fluorescence is proportional to the DNA content in cell. Cells in G1, S and G2 cell cycle phases are differentiated on the basis of different total comet fluorescence. The versatility of application of the comet assay indicates its usefulness in addressing a wide range of questions in biology, medicine and toxicology.

## Advantages of Comet Assay

- Single cell data (distribution data)
- Small cell sample (<10,000 cells)
- Economical, simple and fast
- Sensitivity (<5 cGy gamma rays)
- Any eukaryotic cell

## Disadvantages of Comet Assay

- Single Cell Data (rate limiting)
- Small Cell Sample (sample bias)
- Sensitivity (technical variability)
- Interpretation

# 6

# Micronuclei Induction Assay in Primary Cell Culture

*Atul Prakash and Rajesh Mandil*

Genotoxicity potential of any environmental pollutants can be investigated using the micronucleus (MN) test. It is among the most sensitive DNA damage indicators and it has been applied to several organisms and tissues for evaluation of environmental contaminants (Lemos *et al*., 2011). Micronuclei (MN) assay is more rapid and simpler than chromosomal analysis (Gandhi *et al*., 2003). Compared to the chromosome aberrations, the final result is relatively easily scored and thus requires less time to make an assessment of the clastogenicity of a chemical (Garriott *et a*l. 2002). An *in vitro* MN assay can detect the clastogens and aneugens as well as mitotic delay, apoptosis, chromosome breakage, chromosome loss and non-disjunction (Corvi *et al*., 2008). Although, micronuclei constitute well-characterized biomarkers of chromosomal damage (Salazar *et al.* 2009) but *in vitro* MN test does not provide information on the origin of micronuclei (Corvi *et al*., 2008).

## Experimental Procedure

Micronuclei assay is carried out in PMNC and splenocytes as per method of Hayashi *et al*. (1983).

## Preparation of Chemicals and Reagents

1. **Hank's balance salt solution (HBSS; pH7.2):** 1 g of bovine serum albumin (BSA) and 150 mg of EDTA are dissolved in 100 ml of HBSS of Sigma and pH adjusted to 7.2.
2. **Sorensen's buffer (1/15 M):** 1.82 g of $KH_2PO4$ and 2.36 g of $Na_2HPO4$ are dissolved in 200 ml of distilled water and pH is adjusted to 6.8.
3. **Acridine orange (A.O., pH 7.2):** 12 mg of AO is dissolved in 100 ml of Sorensen's buffer and pH is adjusted to 7.2.

## Procedure

1. Preparation of primary cells culture of different cells namely PMNC and splenocytes form Wistar rats.
2 $10^6$ cells/ml of primary cell culture is used for seeding in 24 well culture plates containing xenobiotics at different μM concentration.
3. Cell culture plates are incubated in into $CO_2$ incubator at 37°C with 5% $CO_2$ for 12 hrs.
4. After incubation period, samples are collected in 1.5 ml micro centrifuge tube and centrifuged at 3200 rpm for 10 min and supernatant is discarded.
5. Cells pellet is dissolve in 1.0 ml of HBSS solution of pH 7.2 and centrifuged for 10 min at 3200 rpm.
6. Supernatant is removed and cells in the suspension are mixed carefully in 100 μl of HBSS.
7. A drop of cell suspension is taken on grease-free clean glass slides and then smeared and stained. Smear should be air dried and fixed with absolute methanol (100 %) for 5 min. Slides are stained with acridine orange for 1 min at room temperature.
8. The slides are rinsed in Sorensen's buffer (pH 6.8) three times and each time slide is kept in Sorensen's buffer for at least three min.
9. Observation of the slide is made on the same day. For micronuclei assay, 1000 cells both mononuclear and binucleated per slide is scored under green fluoresces of fluorescent microscope to determined MN frequencies.

# 7

# Protocol for TUNEL Assay in Primary Cell Culture

*Rajesh Mandil and Atul Prakash*

Terminal deoxynucleotidyl transferase (dUTP) nick end labeling (TUNEL) is a method for detecting DNA fragmentation by labeling the terminal end of nucleic acids. TUNEL is a common method for detecting DNA fragmentation that results from apoptotic signaling cascades. The assay relies on the presence of nicks in DNA which can be identified by terminal deoxynucleotidyl transferase or TdT, an enzyme that will catalyze the addition of dUTPs that are secondarily labeled with a marker. Here we perform the apoptotic assay test by TUNEL assay kit to identify the apoptotic property of any agent or chemical after treatment *of* primary cells culture at different concentration as per protocol provided with TUNEL Assay Kit and apoptotic cells are identified by using fluorescent microscope. Detailed procedure of TUNEL assay is as follows:

1. Preparation of primary cells culture as described in earlier methods.
2. Induction of apoptosis in primary cells culture is done by treatment of primary cells culture with different concentration of xenobiotics and incubation in $CO_2$ incubator for 12 h as mentioned in micronuclei assay method.
3. The positive and negative control cells provided with the kit are already fixed. However, negative control cells are prepared from control sample with absence of inducing agent.
4. Cell Fixation: Cell fixation is a required step in the TUNEL assay and it is done according to the procedure mentioned in the supplied kit by using paraformaldehyde solution. Suspend $1–2 \times 10^6$ cells (treated) in 0.5 ml of phosphate-buffered saline (PBS) and the cell suspension into 5 ml of 1% (w/v) paraformaldehyde in PBS and place on ice for 15 minutes.
5. Centrifuge the cells for 5 minutes at $300 \times g$ and discard the supernatant.

6. Wash the cells in 5 mL of PBS, then pellet the cells by centrifugation as described earlier in step 5.
7. Resuspend the cells in 0.5 ml of PBS.
8. Add the cells to 5 ml of ice-cold 70% (v/v) ethanol. Allow the storage of cells at –20°C in 70% (v/v) ethanol for at least 12–18 hours prior to performing the TUNEL assay for best result. Cells can be stored at –20°C for several days before use.
9. Resuspend the positive (brown cap) and negative (white cap) control cells by swirling the vials. Remove 1 ml aliquot of the control cell suspensions (approximately $1 \times 10^6$ cells/ml). Centrifuge (300 × g) the control cell suspensions for 5 minutes and remove the 70% (v/v) ethanol by aspiration, being careful not to disturb the cell pellet.
10. Resuspend the control cells of each tube with 1 ml of wash buffer (blue cap). Centrifuge for 5 minutes at 300 × g and remove the supernatant by aspiration. Repeat.
11. Prepare DNA-labeling solution: A total volume of 50 µl is required for each sample. Mix 10 µl of reaction buffer (green cap), 0.75 µl of TdT enzyme (yellow cap), 8.0 µl of BrdUTP (violet cap) and 31.25 µl of distilled water. For additional samples, the volumes may be scaled up accordingly, but mix only enough DNA-labeling solution to complete the number of assays prepared per session. The DNA-labeling solution is active for approximately 24 hours.
12. Resuspend the control cell pellets of each tube in 50 µL of the DNA-labeling solution (prepared in step 11).
13. Incubate the cells in DNA-labeling solution for 60 minutes at 37°C in a temperature controlled bath. Shake the samples every 15 minutes to keep the cells in suspension. For samples other than the control cells provided in the kit, incubation times at 37°C may need to be adjusted to longer or shorter periods depending on the characteristics of the experimental samples. DNA-labeling reaction for control cells can also be carried out at 22–24°C overnight.
14. At the end of the incubation time, add 1.0 ml of rinse buffer (red cap) to each tube and centrifuge at 300 × g for 5 minutes. Remove the supernatants by aspiration.
15. Repeat the cell rinsing (as in step 13) with 1.0 ml of rinse buffer (red cap). Centrifuge the samples at 300 x g and remove the supernatants by aspiration.

16. Prepare 100 µl of antibody staining solution for each sample by mixing 5.0 µl of the Alexa FluorR 488 dye– labeled anti-BrdU antibody (orange cap) with 95 µl of rinse buffer (red cap). Prepare only enough antibody staining solution to complete the number of assays prepared per session.
17. Resuspend the cell pellets in 100 µl of the antibody solution prepared in step 16. Incubate the cells in this solution for 30 minutes at room temperature. Protect the samples from light during the incubation.
18. For microscopy applications, it is recommended that the cells be deposited onto slides after the antibody staining step, but prior to the propidium iodide/RNase treatment. Cells that have undergone apoptosis should fluorescence brightly when viewed with blue filter set appropriate for fluorescent.

# 8

# Screening of Myometrial Relaxants in Experimental Dysmenorrhoea an *in vitro* Model

***Afroz Jahan, M.K. Lonare, Sanweer Khatoon and G.S. Rao***

Spontaneous contractions of the uterus are controlled and coordinated for various reproductive functions. The non-pregnant uterus facilitates ova and sperm transport, provides adequate embryo placement for implantation, and contributes to the expulsion of menstrual debris through increased contractility, or periods of relative quiescence (Hutchings *et al.*, 2009). Uterine dysmotility has been implicated in the pathogenesis of certain infertilities, implantation failure, endometriosis, dysmenorrhea, and abnormal menstrual events (Kunz and Leyendecker, 2002; Aguilar and Mitchell, 2010). Although these conditions may not contribute to mortality but can significantly affect patient's reproductive health, and considerably alter their quality of life (Jones *et al.*, 2004; Weissman *et al.*, 2004; Chachamovich *et al.*, 2010). Dysmenorrhoea refers to the occurrence of painful cramps in the lower abdominal region during menstruation and is one of the unresolved gynecological disorders in adolescent girls and women of reproductive age (Wang *et al.*, 2004; Harel, 2008). This condition is mainly characterized by cramping pain in the lower abdomen immediately before or during menstruation, which does affect the quality of their life and work (Davis and Westhoff, 2001).

It is generally accepted that primary dysmenorrhea is a repercussion of decreased uterine blood flow caused by the hyper contractility of the uterine smooth muscle and local contractions of the uterine vessels (Harel *et al.*, 2008). Till date, the cause of dysmenorrhea remains unclear, but it is reported that prostaglandin E2 and F2α are important factors of the regulation of uterine contractility and the levels of both prostaglandins are highest during the first two days of menses when symptoms peak (Andrew and Coco, 1999). Previous etiological studies have demonstrated that close relationship between primary dysmenorrhea (PD) and abnormal increased prostanoid (prostaglandins) secretion (Chan and Hill, 1978; Dawood and Khan-Dawood,

2007). The abnormal prostanoid levels induce frequent or dysrhythmic uterine contractions, which can reduce the uterine blood flow and has been regarded as the main factors leading to menstrual pain (Altunyurt *et al.*, 2005). In addition, clinical reports suggested that increased oxytocin receptor (OTR) expression do occur in PD patients. Principal pharmacological modalities for PD include non-steroidal anti-inflammatory drugs (NSAIDs) or oral contraceptive pills (Sugumar *et al.*, 2013; Witt *et al.*, 2013; Marjoribanks *et al.*, 2015) (Figure 1).

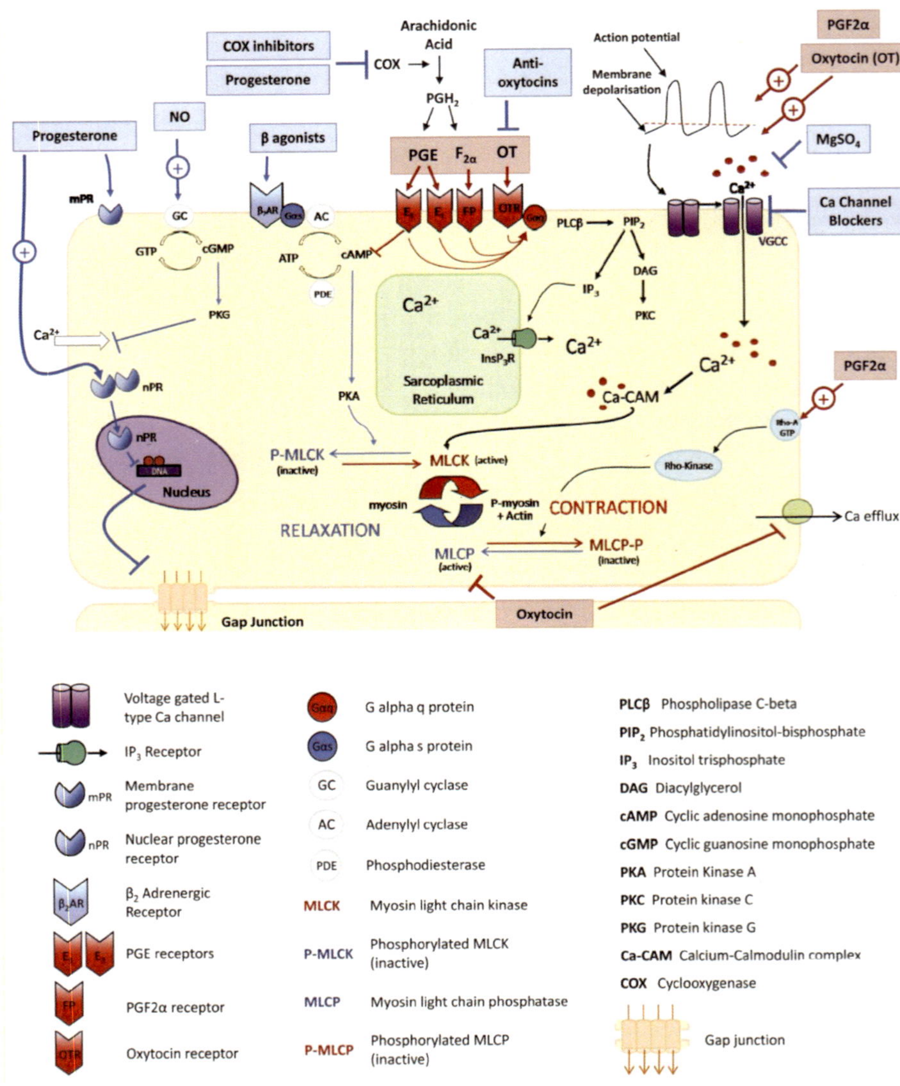


**Figure 1:** Schematic diagram showing major pharmacological pathways to modulate myometrial contraction (https://doi.org/10.1016/j.ogrm.2010.05.001)

However, due to the side effects and lack of satisfactory potency, several tocolytic agents, like cyclooxygenase inhibitors, calcium channel blockers, nitric oxide donors and oxytocin receptor antagonists did not get approval by the US Food and Drug Administration (Haas *et al.*, 2014). To facilitate to screen the uterine smooth muscle relaxants for dysmenorrhea conditions an in vitro mechanistic model is established that has been presented here.

## Therapeutic Significance

Dysmenorrhea being the leading and unresolved cause of gynecological morbidity in women of reproductive age affects from 20 to 93% of this age group and is a leading cause of absenteeism from work. It accounts for over 600 million hours loss from work each year. Because of social taboos females feel discomfort in discussing such conditions and fail to get treated in time. Therefore, it is the need of the hour to address the health condition of women fraternity so that indirectly the economy of India can be supported in terms of minimizing the loss from working hours. Various myometrial relaxants can be screened by this protocol and this procedure will provide insight into the basic data and mechanisms for development of commercial formulations to mitigate inflammation-mediated dysmenorrhea and other gestational disorders.

## Requirements

**Animals:** Rats/ Mice

**Chemicals:** Oestradiol benzoate, Oxytocin, PGF2α Ritodrine, indomethacin, histamine, acetylcholine, $CaCl_2$, EGTA, KCl, Bay K8644 ($CaCl_2$ agonist), salbutamol, L- NAME (NOS inhibitor), 4-Aminopyridine ($K_v$, $K_{ca}$ Channel blocker), glibenclamide ($K_{ATP}$ channel blocker), nifedipine (L-type calcium channel blocker), Y-27632 (Rho kinase inhibitor), TEA ($BK_{Ca}$) blocker), GF-109203X (Protein Kinase C inhibitor), U-73122 (PLC inhibitor), NaCl, sodium bicarbonate ($NaHCO_3$), anhydrous dextrose, potassium dihydrogen orthophosphate ($KH_2PO_4$), magnesium sulphate heptahydrate ($Mg_2SO_4.7H_2O$). All the chemicals should be of high purity and analytical grade.

**Instrumentation:** Isolated organ bath (Digital polygraph with digital data acquisition system, isometric force transducer, 20 ml tissue bath chamber, tissue holder connected to an aerator, stage micrometer to provide tension and water bath connected to a stirrer and a heater), digital pH meter, Millipore water purification system, magnetic stirrer.

## Preparation of Drug Solution

Nifedipine and glibenclamide has to be prepared in ethanol. Oestradiol benzoate for priming the animals should be prepared in absolute alcohol and

0.9% normal saline in 3:7 ratios. All the reagents should be prepared in triple distilled water.

**Composition of various physiological salt solutions for 1 litre**

**Modified Kreb's Henseleit solution (MKHS; pH 7.4)**

| Chemicals | Quantity (g) |
|---|---|
| $NaHCO_3$ | 1.0 |
| NaCl | 6.9 |
| KCl | 0.35 |
| $MgSO_4.7H_2O$ | 0.296 |
| $KH_2PO_4$ | 0.163 |
| d- Glucose | 2.0 |
| $CaCl_2.H_2O$ | 0.367 |

**80 mM Potassium depolarizing solution (KDS; pH 7.4)**

| Chemicals | Quantity (g) |
|---|---|
| $NaHCO_3$ | 1.0 |
| NaCl | 2.566 |
| KCl | 5.874 |
| $MgSO_4.7H_2O$ | 0.296 |
| $KH_2PO_4$ | 0.163 |
| d- Glucose | 2.0 |
| $CaCl_2.H_2O$ | 0.367 |

**$Ca^{2+}$ free MHKS with EGTA solution- pH 7.4**

| Chemicals | Quantity (g) |
|---|---|
| $NaHCO_3$ | 1.0 |
| NaCl | 6.9 |
| KCl | 0.35 |
| $MgSO_4.7H_2O$ | 0.296 |
| $KH_2PO_4$ | 0.163 |
| d- Glucose | 2.0 |
| EGTA | 0.380 |

**$Ca^{2+}$ free KDS without EGTA; pH 7.4**

| Chemicals | Quantity (g) |
|---|---|
| $NaHCO_3$ | 1.0 |

| | |
|---|---|
| NaCl | 2.566 |
| KCl | 5.874 |
| $MgSO_4.7H_2O$ | 0.296 |
| $KH_2PO_4$ | 0.163 |
| d-Glucose | 2.0 |

**$K^+$ free MHKS solution; pH 7.4**

| **Chemicals** | **Quantity (g)** |
|---|---|
| $NaHCO_3$ | 1.0 |
| NaCl | 7.17 |
| $MgSO_4.7H_2O$ | 0.296 |
| $NaH_2PO_4$ | 0.165 |
| d- Glucose | 2.0 |
| $CaCl_2.H_2O$ | 0.367 |

## Induction of Dysmenorrhea Model

Experimental dysmenorrhea can be induced as per the protocol described by Ostad *et al.* (2001) and Yang *et al.* (2015). Briefly, divide the total female animals into two groups. Group I, female mice pre-treated with estradiol benzoate (1mg/kg/d) subcutaneously for three consecutive days. Group II, female mice injected with normal saline s/c will be served as control group. On the fourth day, mice injected with 0.4 U of oxytocin intra-peritoneally. Next, the writhing responses, which mainly consisted of abdominal wall contractions, pelvic rotation, and followed by hind limb stretches, will be observed and recorded for 30 min and percent inhibition of writhing response (Figure 2) can be calculated (Doherty *et al.* 1987).

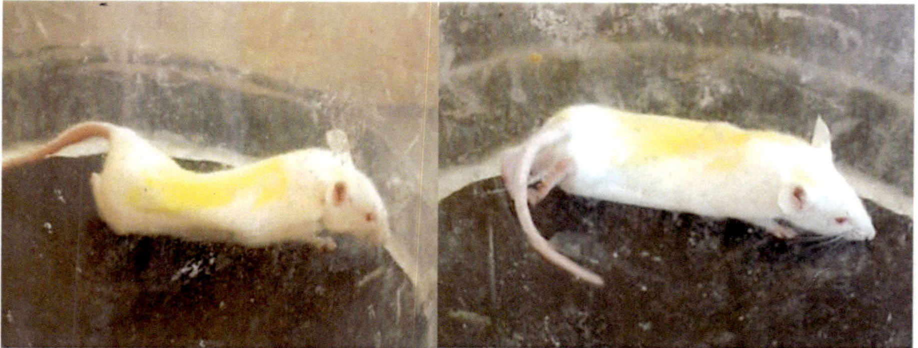

**Figure 2:** Writhing responses in experimental dysmenorrhea mouse model showing abdominal wall contractions and hind limb stretches

## Isolation and Mounting of Myometrium and Recording of Isometric Tension

1. Sacrificed the mice as per the CCSEA (Committee for Control and Supervision of Experiments on Animals) standard procedure by cervical dislocation after anesthetizing the animals with Urethane (1200 mg/kg) or Ketamine (80 mg/kg, intra peritoneal) and Xylazine (16 mg/ kg, intra peritoneal).
2. Uterine horns dissected out and transferred to a wash containing Kreb's solution. Cuts the longitudinal uterine stripes after separating the adjacent fascia.
3. Then mount the tissue in stainless steel hook attached to the S-shaped notch in oxygen delivery tube and suspended inner organ bath / chamber containing 20 ml modified Krebs-Henseleit solution.
4. The other end of the uterine strip tied to an isometric force transducer.
5. The bath temperature has to be maintained at 37.0± 0.5°C and aerated continuously with carbogen (95% $O_2$ and ~5% $CO_2$ ).
6. The tension will be recorded using a polygraph digital data acquisition system linked to isometric transducer connected to a recorder.
7. The uterine strip mounts under a preload resting tension of 1gram and allowed to equilibrate for about 60 minutes.
8. During the equilibration period, wash thrice with modified Krebs-Henseleit solution for every 15 minutes before proceeding for recording of dose response.
9. Further, contractions and relaxations can be recorded using a polygraph digital data acquisition system linked to isometric transducer connected to a recorder.

## Recording of Drug Responses on Myometrium

Before start of recording of various drug responses on myometrium the vitality of tissue has to be checked by adding 80 mM KCl to the tissue bath. If the tissue is viable, it will contract due to opening of voltage dependant $Ca^{2+}$ channels. The uterine strips are treated with established inhibitors or blockers of various pathways or receptors to characterize the mechanism involved in tocolytic activity.

## 1. Recording of Spontaneous Myogenic Activity of Myometrium

After mounting the myometrium in isolated tissue bath, tissues should be allowed to equilibrate and spontaneous rhythmic contractions of myometrium are recorded to determine the amplitude and frequency of myogenic activity.

## 2. Recording of Contractions Induced by Receptor Mediated Spasmogens

Cumulative contractile dose response curve can be obtained with graded doses of oxytocin ($1.10 \times 10^{-13}$ to $3.53 \times 10^{-12}$M), $PGF_{2\alpha}$ ($8.72 \times 10^{-9}$ to $2.79 \times 10^{-7}$M), histamine ($10^{-8}$ to $10^{-3}$M) and acetylcholine ($1 \times 10^{-7}$ to $3 \times 10^{-4}$M).After recording the response of each drug, sufficient washing should be given to the tissue so that it regains its normal spontaneity before proceeding for next tissue response.

## 3. Recording of Contractions Induced by Ion Channel Mediated Spasmogens

Cumulative contractile dose response curve can be obtained with graded doses of $CaCl_2$ (10μM to 10mM) in $Ca^{2+}$ free MKHS with and without EGTA respectively. In this protocol, myometrial strips are initially exposed to $Ca^{2+}$-free MKHS solution containing 3 mM EGTA and after two successive washings with this solution, tissues are incubated with $Ca^{2+}$-free high depolarizing solution (to open the L-type $Ca^{2+}$ channels) before recording the dose response of calcium chloride alone. Effect of single dose of KCl (80mM) alone and Bay K8644 (0.1 nM to 10 μM) can be obtained. After recording the response of each drug, give the sufficient washing to the tissue so that it regains its normal spontaneity before proceeding for next tissue response.

## 4. Recording of Myometrial Relaxant Activity

i. **Effect of relaxants on oxytocin, $PGF_{2\alpha}$,KCl induced contractions:** The myometrium is pre-contracted with submaximal concentration of oxytocin ($1.76 \times 10^{-12}$M), $PGF_{2\alpha}$ ($1.34 \times 10^{-7}$M), KCl (80mM) and relaxed with cumulative doses of Ritodrine ($0.5 \times 10^{-13}$ to $1.5 \times 10^{-9}$M), Salbutamol ($10^{-13}$ to $10^{-5}$M) and L- NAME ($10^{-9}$ to $10^{-5}$M).

ii. **Effect on spasmogens after pre-incubation with relaxants:** The myometrium is pre-incubated with indomethacin (60μM), 4-Amino-pyridine (1mM), glibenclamide (10μM), nifedipine (1μM), L- NAME (0.1mM), Y-27632 (10μM), TEA (10 mM), GF-109203X (1 μM) and U-73122 (10 μM) for 30 minutes. After incubation cumulative contractile dose response of oxytocin and $PGF_{2\alpha}$ is recorded.

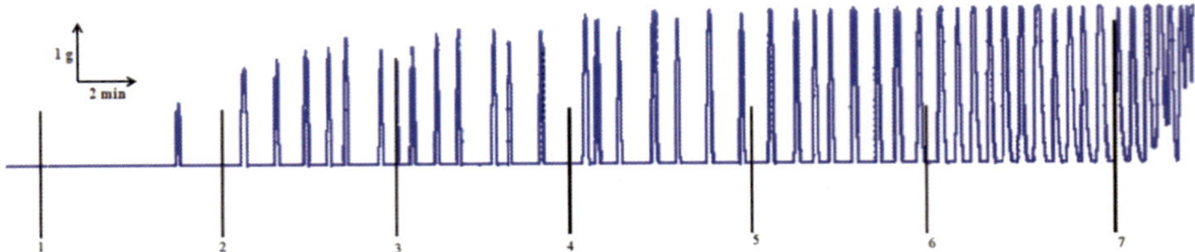


**Figure 3:** Representative physiographic recording of Oxytocin cumulative contractile dose (0.11x10-12to 3.53x10-12M) response in myometrium of dysmenorrhoeic mice.

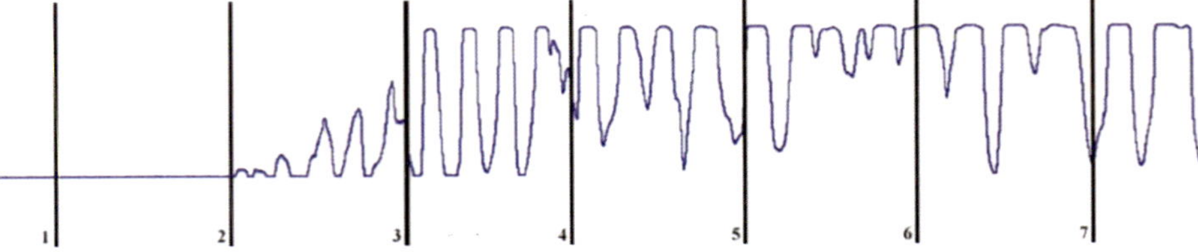


**Figure 4:** Representative physiographic recording of PGF2α cumulative contractile dose (8.72x10$^{-9}$to 2.79x10$^{-7}$M) response in myometrium of dysmenorrhoeic mice.

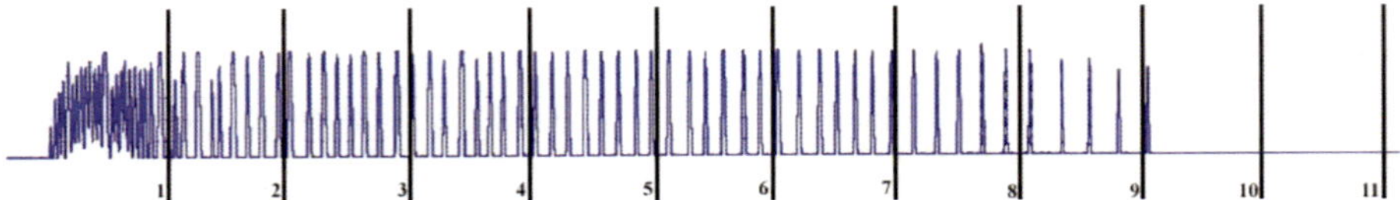


**Figure 5:** Representative physiographic recording of cumulative dose dependent relaxation response of ritodrine (0.5x10$^{-13}$to 1.5x10$^{-9}$M) after pre-contracted with Oxytocin (1.76x10$^{-12}$M) in myometrium of dysmenorrhoeic mice.

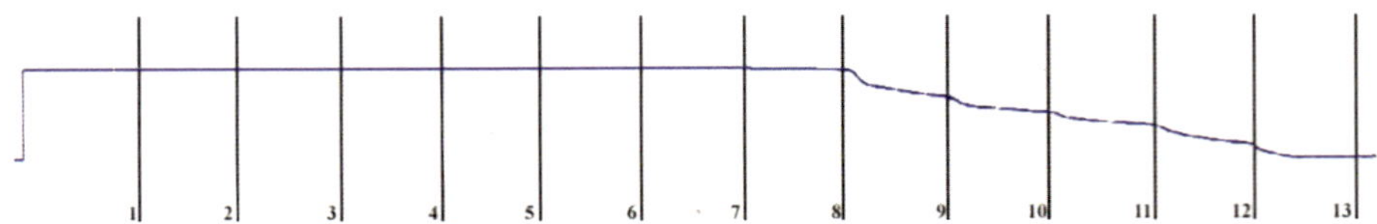


**Figure 6:** Representative physiographic recording of cumulative dose dependent relaxation response of ritodrine (0.5x10$^{-13}$to 1.5x10$^{-9}$M) after pre-contracted with KCl (80mM) in myometrium of dysmenorrhoeic mice.

## Data Analysis

The results can be graphically represented as Mean±SEM. The log concentration producing 50% of the maximal effect (EC-50 for agonist and IC-50 for antagonists) and $pD_2$ values are calculated for each agent using computer assisted non- linear regression analysis with variable slope (GraphPad Prism, San Diego, CA) (Prism, 2009). The EC-50 and IC-50 values can be presented as geometric means accompanied by their respective 95% confidence limits. The differences between the multiple groups can be analyzed using two way

ANOVA followed by Dunnett or Bonoferroni post hoc test. A value of $P< 0.05$ is considered significant.

## References

Aguilar H N and Mitchell B F (2010) Physiological pathways and molecular mechanisms regulating uterine contractility. Human Reproduction Update 16(6): 725-744.

Altunyurt S, Gol M, Altunyurt S, Sezer O and Demir N (2005) Primary dysmenorrhea and uterine blood flow: a color Doppler study. Journal of Reproductive Medicine50(4): 251-255.

Andrew S and Coco M D (1999) Primary dysmenorrhoea. Am Fam Physician 60: 489-496.

Chachamovich J R, Chachamovich E, Ezer H, Fleck M P, Knauth D and Passos E P (2010) Investigating quality of life and health-related quality of life in infertility: a systematic review. Journal of Psychosomatic Obstetrics and Gynecology31(2): 101-110.

Chan W Y and Hill J C (1978) Determination of menstrual prostaglandin levels in non-dysmenorrheic and dysmenorrheic subjects. Prostaglandins15(2): 365-375.

Davis A R and Westhoff C L (2001) Primary dysmenorrhea in adolescent girls and treatment with oral contraceptives. Journal of Pediatric and Adolescent Gynecology 14(1): 3-8.

Dawood M Y and Khan-Dawood F S (2007) Clinical efficacy and differential inhibition of menstrual fluid prostaglandin F2 alpha in a randomized, double-blind, crossover treatment with placebo, acetaminophen and ibuprofen in primary dysmenorrhea. American Journal of Obstetrics and Gynecology 196(1): 35.e1-35.e5.

Doherty N S, Beaver T H, Chan K Y, Coutant J E and Westrich G L (1987) The role ofprostaglandins in the nociceptive response induced byintra-peritoneal injection of zymosan in mice. Britwash Journal of Pharmacology91(1): 39-47.

Ghosh MN. (2005). In: Fundamentals of Experimental Pharmacology. 3rd Edn

Haas D M, Benjamin T, Sawyer R, Quinney S K (2014) Short-term tocolytics for preterm delivery - current perspectives. Int. J. Womens Health 6, 343-349.

Harel Z (2008) Dysmenorrhea in adolescents. Annals of the New York Academy of Sciences 1135(1): 185-195.

Hutchings G, Williams O, Cretoiu D and Ciontea S M (2009) Myometrial interstitial cells and the coordination of myometrial contractility. Journal of Cellular and Molecular Medicine 13(10): 4268-4282.

Jones G, Jenkinson C and Kennedy S (2004) The impact of endometriosis upon quality of life: a qualitative analysis. Journal of Psychosomatic Obstetrics &Gynecology25(2): 123-133.

Kunz G and Leyendecker G (2002) Uterine peristaltic activity during the menstrual cycle: characterization, regulation, function and dysfunction. Reproductive biomedicine Online 4(3): 5-9.

Marjoribanks J, Ayeleke R O, Farquhar C and Proctor M (2015) Non-steroidal anti-inflammatory drugs for dysmenorrhoea. Cochrane Database of Systematic Review 30(7): CD001751.

Ostad S N, Soodi M, Shariffzadeh M, Khorshidi N and Marzban H (2001) The effect of fennel essential oil on uterine contraction as a model for dysmenorrhea, pharmacology and toxicology study. Journal of Ethnopharmacology 76(3): 299-304.

Prism G 7.02 (2009) GraphPad Software, San Diego, CA, USA.

Sugumar R, Krwashnaiah V, Channaveera G S and Mruthyunjaya S (2013) Comparison of the pattern, efficacy, and tolerability of self-medicated drugs in primary dysmenorrhea: a questionnaire based survey. Indian Journal of Pharmacology 45(2): 180-183.

Wang L, Wang X, Wang W, Chen C, Ronnennberg A G, Guang W, Huang A, Fang Z, Zang T, Wang L and Xu X (2004) Stress and dysmenorrhoea: a population based prospective study. Occupational and Environmental Medicine 61(12): 1021-1026.

Weissman A M, Hartz A J, Hansen M D and Johnson S R (2004) The natural history of primary dysmenorrhoea: a longitudinal study BJOG: An International Journal of Obstetrics & Gynaecology 111(4): 345-352.

Witt J, Strickland J, Cheng A L, Curtis C and Calkins J (2013) A randomized trial comparing the VIPON tampon and ibuprofen for dysmenorrhea pain relief. Journal of Women's Health 22(8): 702-705.

Yang L, Cao Z, Yu B and Chai M D (2015) An in vivo mouse model of primary dysmenorrhea. Experimental Animals 64(3): 295-303.

# 9

# Direct and Indirect LPS Induced Acute Lung Injury Model

*Meemansha Sharma, Anshuk Sharma, Thakur Uttam Singh*

## Introduction

Acute lung injury (ALI) is the manifestation of an inflammatory response of lung resulting from severe sepsis, bacterial or viral pneumonia, trauma, and burns in animals (Ali *et al.,* 2020).It is distinguwashed by disruption to the alveolar capillary membranes, atelectasis of the pulmonary airspaces, and neutrophil and protein-rich fluid infiltration into the alveolar space (Li *et al*., 2019). Although advances in treatment, the disability and fatality rates of ALI remain high, that is around 30-40% (Ali *et al.,* 2020). At the moment, new therapeutic alternatives are urgently required to treat ALI. As a result, this is a very active area of research to understand the molecular principles underlying pulmonary inflammation and ALI (Reiss *et al.,* 2012). The use of animal models of inflammatory lung has allowed for the probing of pathways of inflammation responsible for injury. Specially, rodent models may serve a link between in vitro research to large animals and clinical studies by allowing for high quantities of samples under physiological circumstances at low expenses (Domscheit *et al.,* 2020).LPS, found in the outer membrane of gram-negative bacteria, consists of a polar lipid head group (lipid A) and a chain of repeating disaccharides (Gao *et al.,* 2023). LPS induced lung injury model, closely mimic sepsis and inflammatory lung diseases, is a well-establwashed model in this field of research (Reiss *et al.,* 2012). To mimic acute lung injury, LPS can be administered either directly through intratracheal or intranasal route (direct pulmonary insult) or indirectly through intraperitoneally and intravenous routes (indirect extra pulmonary insult) (Perl *et al.,* 2011). This chapter focuses mostly on rodent model of acute lung injury caused by lipopolysaccharide (LPS), both direct and indirect. Additionally, it also includes the characteristics and underlying molecular mechanism of LPS induced lung injury along with applications in drug development.

## Characteristics of LPS-induced Acute Lung Injury in Different Animal Models

The ideal animal model represents of an immediate inflammatory response with increased permeability epithelial cells of alveoli, resulting neutrophil influx, and protein rich alveolar exudates accumulation in the alveolar tissue of lungs following stimulus with LPS, inducing acute lung injury (Wiscombe, 2021).Large animal models, such as primates, dogs, sheep, or pigs, are the only ones that allow experiments in ventilated animals over an extended period of time (Chen *et al.,* 2010).Due to the requirement for an animal intensive care centre and the scarcity of molecular reagents for large animals, these models are quite costly. Small animal research models, such as mice, rats, and rabbits, have demonstrated value in examining particular molecular pathways, but the ability to extrapolate results to humans is restricted. Presence of PIM (pulmonary intravascular macrophages) plays important role in development of endotoxin induced lung injury among different animal species (Gill *et al.,* 2008). Species such as sheep, goat, pig, cattle and horse, have PIM in pulmonary vasculature so that LPS and particulate matter localized in lungs and enhance the susceptibility (required dose of LPS in µg/kg range) towards the acute lung injury (Sipahi and Atalay, 2014). In contrast, dog, rabbits, rodents, primates and humans have less PIM and intravascular particulate localized in liver and spleen and required very high dose of LPS (in mg/kg range) to manifest lung injury (Sipahi and Atalay, 2014). Clinical investigations and animal models indicate the involvement of nitric oxide (NO) in the development of lung injury (Afshari *et al.,*2011; Liu *et al.,* 2016).Noteworthy variations in NO pathways have been documented, showing species-specific differences (Matute-Bello *et al.,* 2008). Further research indicates substantial distinctions in the capacity for nitric oxide (NO) production between human and rodent macrophages. Rodent macrophages exhibit a high production of NO, while normal, non-activated human macrophages generate minimal amounts of NO (Suleimanov *et al.,* 2024). In comparison, cattle macrophages produce NO, whereas macrophages from hamsters, monkeys, goats, and pigs resemble human macrophages in their low NO generation (Matute-Bello *et al.,* 2008).Interleukins such as IL-8 serves as a potent chemotactic factor for neutrophils and plays an active role in the pathogenesis of acute lung injury (ALI) (Allen &Kurdowska, 2014).In response to bacterial products or endotoxin, IL-8 and its related chemokines are produced in various species. Despite the absence of a CXCL8 gene in rats and mice, they generate specific chemokines, including macrophage inflammatory protein-2 (MIP-2) (Sipahi & Atalay, 2014).

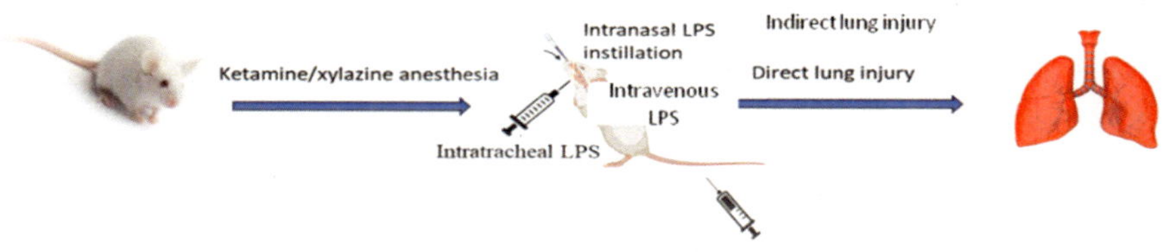


**Figure 1:** Depicts direct and indirect routes of LPS administration to induce acute lung injury in mouse

## Direct and Indirect Administration of Lipopolysaccharide Induces Lung Injury in Rodents

LPS-induced lung injury is one of the most commonly used rodent models for ARDS (Shaaban *et al.,* 2018). The outer membrane of Gram-negative bacteria hosts lipopolysaccharide, also termed endotoxins, that can mimic sepsis and ALI to a great extent (Zou *et al.,* 2017). Experimental evidence proposed that different pathophysiological pathways are activated during pulmonary and extra-pulmonary LPS challenge, especially during the early phase of disease progression (Hu *et al.,* 2013). Direct lung injury can be modelled in rodents by administration of LPS to the lungs through either tracheal instillation or inhalation (Reiss *et al.,* 2012). In this case, the alveolar epithelium is the primary structure that is damaged. Local administration of LPS causes an acute and vigorous migration of inflammatory cells and pro-inflammatory cytokines into the lung tissue with resolution by 72 h after the exposure followed by secondary fibrosis. Notably, mediators like TNF-α reach their highest levels within a few hours after the administration of endotoxin, preceding the migration of neutrophils (Kumari *et al.,* 2017). Intravenous or intraperitoneal LPS administration triggers the release of inflammatory mediators into the systemic circulation, which in turn evokes indirect lung injury (Keskinidov *et al.,* 2022). In this case, the pulmonary vascular endothelium is the primary structure that is damaged and interstitial oedema is the most prominent pathophysiological alteration. In the pulmonary system, introducing LPS, whether through intravenous or intra-alveolar means, leads to common alterations in polymorphonuclear (PMN) cells deformability and the capture of PMN in the pulmonary capillaries (Grommes and Soehnlein, 2011). However, the instillation of intratracheal LPS results in significant elevations of PMN in the air spaces as compared to intravenous routes. Fig 1 shows direct and indirect routes of LPS administration to induce acute lung injury in mouse.

## Signailing Pathways Involved in Acute Lung Injury Induced by Lipopolysaccharide Administration

Endotoxins activate the innate immune response through their activation of TLR4 receptors. LPS bind to TLR4 after interacting with extracellular binding proteins such as CD14 and MD-2 (Ciesielska *et al.,* 2021). Further, Intracellular adaptor proteins (MyD88) link to the TLR-4 receptor and initiate activation of interleukin -1R- associated kinases (IRAK). When IRAK-1 is phosphorylated, it triggers the activation of a protein complex consisting of TRAF6, TAK1, TAB1, and TAB2. This activation, in turn, leads to the initiation of the NF-kB and MAPK pathways, along with the induction of pro-inflammatory cytokines such as IL-1β, TNF-α and IL-6 (Joh *et al.,* 2012).A notable and significant increase in the phosphorylation of extracellular regulatory protein kinases (ERK 1 and 2), p38 mitogen-activated protein kinase (p38 MAPK), and c-Jun amino acid kinases (c-Jun N-terminal kinase, JNK) has been reported following lipopolysaccharide (LPS) stimulation in acute lung injury (ALI) conditions (Li *et al.*, 2022).ERK, p38, and JNK are key regulators of MAPK pathways in mammals (Li *et al.,* 2022).The other signalling pathway involving phosphatidylinositol 3′-kinase (PI3K) and Akt plays a crucial role in regulating both cell survival and responses to oxidative stress during pulmonary inflammation (Meng *et al.,* 2018). Akt has the ability to modulate NF-κB activation and govern the activity of the mammalian target of rapamycin (mTOR) (Jiang *et al.,* 2017)

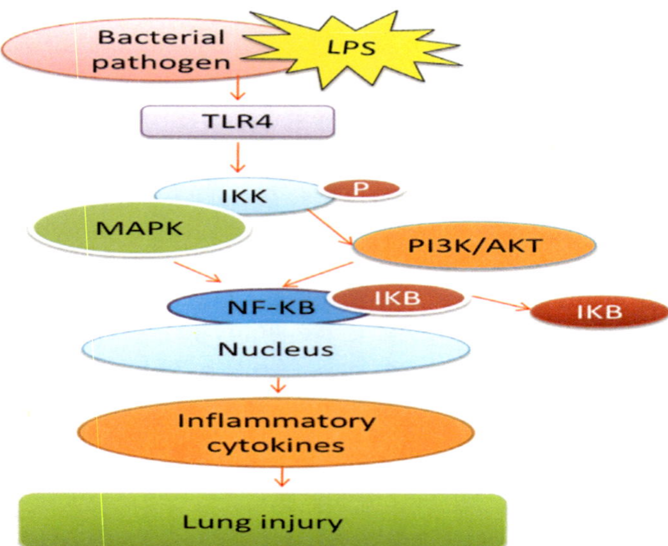


**Figure 2:** Different signaling pathways involve in LPS induced acute lung injury

## Application of LPS-induced Lung Injury Model in Drug Development

The LPS-induced lung injury model has been employed in therapeutic assessments for anti-inflammatory, anti-thrombotic, stem cell, and chemotherapy evaluations. Administering a low dose of dexamethasone has the potential to decrease pulmonary inflammation and fibrosis following LPS-induced acute lung injury (ALI) in rats. Additionally, it may enhance the expression of glucocorticoid receptors in the lung, likely by increasing glucocorticoid receptor levels and facilitating the nuclear translocation of glucocorticoid receptor protein (Wang *et al.,* 2008). Various antibiotics, including florfenicol, tetracycline, and telithromycin, have been demonstrated a decrease release of inflammatory mediators, such as TNF-a, in broncho-alveolar lavage (BAL). This protective outcome is likely associated with the inhibition of NF-kB in LPS induced acute lung injury model *in vitro* and *in vivo* (Chen *et al.,* 2010). ALI is characterized by significant accumulation of fibrin outside blood vessels. Additionally, instances of thrombosis in the pulmonary vasculature and disseminated intravascular coagulation have been observed in association with ARDS (Whyte *et al.,* 2020).Nebulized antithrombin (AT) and heparin mitigate lung injury by reducing both coagulation and inflammation, without affecting systemic coagulation or causing bleeding (Camprubi-Rimblas *et al.,* 2020). Further, MSC (Mesenchymal stem cells) administration significantly reduced LPS-induced pulmonary inflammation, evidenced by decreased total cell and neutrophil counts in BAL, as well as lower proinflammatory cytokine levels in BAL fluid and lung parenchymal homogenates (Pedrazza *et al.,* 2017). In addition, some chemotherapeutic agents have also been evaluated against ALI and had potential to reduce tissue damage. Intravenous taxol reduced ALI-associated inflammation and vascular leakage in vivo (Mirzapoiazova *et al.,* 2007). Pre-treating with cyclosporin A, a potent mitochondrial permeability transition inhibitor, protected against LPS-induced mitochondrial abnormalities. This independent effect contributed to the suppression of LPS-induced TNF-α production (Crouser *et al.,* 2002).

## Pros and Cons Using LPS-induced Lung Injury Model

Using bacterial LPS (lipopolysaccharide) represents several benefits as an approach to simulate the impact of gram-negative bacteria in both animals and humans (D'Aessio, 2018). Administration of LPS is simple and the outcomes typically demonstrate reproducibility within experiments. LPS effectively triggers innate immune responses through the TLR4 pathways and demonstrates minimal direct toxicity to cells in vitro (Fanelli and Ranieri, 2015). Thus, the use of LPS provides information about the effects of host inflammatory responses, which occur in bacterial infections. Nevertheless, there have been several documented drawbacks to using the LPS-induced lung injury model. The purity of LPS preparations varies, and they may contain contaminants such as bacterial lipoproteins and other microbial substances (Matute-Bello *et al.,* 2008).Typically, the administration of LPS does not induce the significant endothelial and epithelial damage seen in humans experiencing acute respiratory distress syndrome(Matute-Bello *et al.,* 2008). Bacteria generate potent cellular toxins known as exotoxins and employ various effector systems, leading to direct cellular lysis (Roy-Burman *et al.,* 2001). Consequently, LPS alone offers an insufficient representation of the impact of live bacteria in the lungs.

## Conclusion

There is increasing information suggesting that LPS induced ALI model is widely used to study the molecular mechanism of inflammatory lung disease and emergence of innovative treatment approaches for animal and humans. For screening and optimizing new drugs in terms of EC50, Dose dependent biological effects and therapeutic window, these animal models are reliable and reproducible. However, there is variation in ALI clinical manifestations in different animal models that can be attributed to heterogenesity in host innate immune responses.Hence, there is need to compare proteins and gene profiling between the LPS-induced ALI model and our target animal species or humans with acute lung injury.

## References

Afshari, A., Brok, J., Møller, A. M., & Wetterslev, J. (2011). Inhaled nitric oxide for acute respiratory distress syndrome and acute lung injury in adults and children: a systematic review with meta-analysis and trial sequential analysis. Anesthesia and analgesia, 112(6), 1411–1421. https://doi.org/10.1213/ANE.0b013e31820bd185

Ali, H., Khan, A., Ali, J., Ullah, H., Khan, A., Ali, H., Irshad, N., & Khan, S. (2020). Attenuation of LPS-induced acute lung injury by continentalic acid in rodents through inhibition of inflammatory mediators correlates with increased Nrf2 protein expression. BMC pharmacology & toxicology, 21(1), 81. https://doi.org/10.1186/s40360-020-00458-7

Allen, T. C., &Kurdowska, A. (2014). Interleukin 8 and acute lung injury. Archives of pathology & laboratory medicine, 138(2), 266–269. https://doi.org/10.5858/arpa.2013-0182-RA

Camprubí-Rimblas, M., Tantinyà, N., Guillamat-Prats, R., Bringué, J., Puig, F., Gómez, M. N., Blanch, L., & Artigas, A. (2020). Effects of nebulized antithrombin and heparin on inflammatory and coagulation alterations in an acute lung injury model in rats. Journal of thrombosis and haemostasis : JTH, 18(3), 571–583. https://doi.org/10.1111/jth.14685

Chen, H., Bai, C., & Wang, X. (2010). The value of the lipopolysaccharide-induced acute lung injury model in respiratory medicine. Expert review of respiratory medicine, 4(6), 773–783. https://doi.org/10.1586/ers.10.71

Ciesielska, A., Matyjek, M., & Kwiatkowska, K. (2021). TLR4 and CD14 trafficking and its influence on LPS-induced pro-inflammatory signaling. Cellular and molecular life sciences : CMLS, 78(4), 1233–1261. https://doi.org/10.1007/s00018-020-03656-y

Crouser, E. D., Julian, M. W., Joshi, M. S., Bauer, J. A., Wewers, M. D., Hart, J. M., & Pfeiffer, D. R. (2002). Cyclosporin A ameliorates mitochondrial ultrastructural injury in the ileum during acute endotoxemia. Critical care medicine, 30(12), 2722–2728. https://doi.org/10.1097/00003246-200212000-00017

Domscheit, H., Hegeman, M. A., Carvalho, N., & Spieth, P. M. (2020). Molecular Dynamics of Lipopolysaccharide-Induced Lung Injury in Rodents. Frontiers in physiology, 11, 36. https://doi.org/10.3389/fphys.2020.00036

D'Alessio F. R. (2018). Mouse Models of Acute Lung Injury and ARDS. Methods in molecular biology (Clifton, N.J.), 1809, 341–350. https://doi.org/10.1007/978-1-4939-8570-8_22

Fanelli, V., & Ranieri, V. M. (2015). Mechanisms and clinical consequences of acute lung injury. Annals of the American Thoracic Society, 12 Suppl 1, S3–S8. https://doi.org/10.1513/AnnalsATS.201407-340MG

Gao, Y., Widmalm, G., &Im, W. (2023). Modeling and Simulation of Bacterial Outer Membranes with Lipopolysaccharides and Capsular Polysaccharides. Journal of chemical information and modeling, 63(5), 1592–1601. https://doi.org/10.1021/acs.jcim.3c00072

Gill, S. S., Suri, S. S., Janardhan, K. S., Caldwell, S., Duke, T., & Singh, B. (2008). Role of pulmonary intravascular macrophages in endotoxin-induced lung inflammation and mortality in a rat model. Respiratory research, 9(1), 69. https://doi.org/10.1186/1465-9921-9-69

Grommes, J., & Soehnlein, O. (2011). Contribution of neutrophils to acute lung injury. Molecular medicine (Cambridge, Mass.), 17(3-4), 293–307. https://doi.org/10.2119/molmed.2010.00138

Hu, R., Xu, H., Jiang, H., Zhang, Y., & Sun, Y. (2013). The role of TLR4 in the pathogenesis of indirect acute lung injury. Frontiers in bioscience (Landmark edition), 18(4), 1244–1255. https://doi.org/10.2741/4176.

Jiang, W., Liu, J., Li, P., Lu, Q., Pei, X., Sun, Y., Wang, G., & Hao, K. (2017). Magnesium isoglycyrrhizinate shows hepatoprotective effects in a cyclophosphamide-induced model of hepatic injury. Oncotarget, 8(20), 33252–33264. https://doi.org/10.18632/oncotarget.16629

Joh, E. H., Gu, W., & Kim, D. H. (2012). Echinocystic acid ameliorates lung inflammation in mice and alveolar macrophages by inhibiting the binding of LPS to TLR4 in NF-κB and MAPK pathways. Biochemical pharmacology, 84(3), 331–340. https://doi.org/10.1016/j.bcp.2012.04.020

Keskinidou, C., Vassiliou, A. G., Dimopoulou, I., Kotanidou, A., & Orfanos, S. E. (2022). Mechanistic Understanding of Lung Inflammation: Recent Advances and Emerging Techniques. Journal of inflammation research, 15, 3501–3546. https://doi.org/10.2147/JIR.S282695

Kumari, A., Dash, D., & Singh, R. (2017). Curcumin inhibits lipopolysaccharide (LPS)-induced endotoxemia and airway inflammation through modulation of sequential release of inflammatory mediators (TNF-α and TGF-β1) in murine model. Inflammopharmacology, 25(3), 329–341. https://doi.org/10.1007/s10787-017-0334-3

Li, W. W., Wang, T. Y., Cao, B., Liu, B., Rong, Y. M., Wang, J. J., Wei, F., Wei, L. Q., Chen, H., & Liu, Y. X. (2019). Synergistic protection of matrine and lycopene against lipopolysaccharide-induced acute lung injury in mice. Molecular medicine reports, 20(1), 455–462. https://doi.org/10.3892/mmr.2019.10278

Li, Z., Fu, X., Fan, Y., Zhao, C., Wang, Q., Feng, L.,& Fan, J. (2022). Effect of epicatechin on inflammatory cytokines and MAPK/NF-Î° B signaling pathway in lipopolysaccharide-induced acute lung injury of BALB/c mice. General Physiology & Biophysics, 41(4).

Liu, W. W., Han, C. H., Zhang, P. X., Zheng, J., Liu, K., & Sun, X. J. (2016). Nitric oxide and hyperoxic acute lung injury. Medical gas research, 6(2), 85–95. https://doi.org/10.4103/2045-9912.184718

Matute-Bello, G., Frevert, C. W., & Martin, T. R. (2008). Animal models of acute lung injury. American Journal of Physiology-Lung Cellular and Molecular Physiology, 295(3), L379-L399.

Meng, L., Li, L., Lu, S., Li, K., Su, Z., Wang, Y., Fan, X., Li, X., & Zhao, G. (2018). The protective effect of dexmedetomidine on LPS-induced acute lung injury through the HMGB1-mediated TLR4/NF-κB and PI3K/Akt/mTOR pathways. Molecular immunology, 94, 7–17. https://doi.org/10.1016/j.molimm.2017.12.008

Mirzapoiazova, T., Kolosova, I. A., Moreno, L., Sammani, S., Garcia, J. G., & Verin, A. D. (2007). Suppression of endotoxin-induced inflammation by taxol. The European respiratory journal, 30(3), 429–435. https://doi.org/10.1183/09031936.00154206

Pedrazza, L., Cunha, A. A., Luft, C., Nunes, N. K., Schimitz, F., Gassen, R. B., Breda, R. V., Donadio, M. V., de Souza Wyse, A. T., Pitrez, P. M. C., Rosa, J. L., & de Oliveira, J. R. (2017). Mesenchymal stem cells improve survival in LPS-induced acute lung injury acting through inhibition of NETs formation. Journal of cellular physiology, 232(12), 3552–3564. https://doi.org/10.1002/jcp.25816

Perl, M., Lomas-Neira, J., Venet, F., Chung, C. S., & Ayala, A. (2011). Pathogenesis of indirect (secondary) acute lung injury. Expert review of respiratory medicine, 5(1), 115–126. https://doi.org/10.1586/ers.10.92

Reiss, L. K., Uhlig, U., & Uhlig, S. (2012). Models and mechanisms of acute lung injury caused by direct insults. European journal of cell biology, 91(6-7), 590–601. https://doi.org/10.1016/j.ejcb.2011.11.004

Roy-Burman, A., Savel, R. H., Racine, S., Swanson, B. L., Revadigar, N. S., Fujimoto, J., Sawa, T., Frank, D. W., & Wiener-Kronwash, J. P. (2001). Type III protein secretion is associated with death in lower respiratory and systemic Pseudomonas aeruginosa infections. The Journal of infectious diseases, 183(12), 1767–1774. https://doi.org/10.1086/320737

Shaaban, A. A., El-Kashef, D. H., Hamed, M. F., & El-Agamy, D. S. (2018). Protective effect of pristimerin against LPS-induced acute lung injury in mice. International immunopharmacology, 59, 31–39. https://doi.org/10.1016/j.intimp.2018.03.033

Sipahi, E. Y., & Atalay, F. (2014). Experimental Models of Acute Lung Injury. Eurasian Journal of Pulmonology, 16(2).

Suleimanov, S. K., Efremov, Y. M., Klyucherev, T. O., Salimov, E. L., Ragimov, A. A., Timashev, P. S., & Vlasova, I. I. (2024). Radical-Generating Activity, Phagocytosis, and Mechanical Properties of Four Phenotypes of Human Macrophages. International journal of molecular sciences, 25(3), 1860. https://doi.org/10.3390/ijms25031860

Wang, X. Q., Zhou, X., Zhou, Y., Rong, L., Gao, L., & Xu, W. (2008). Low-dose dexamethasone alleviates lipopolysaccharide-induced acute lung injury in rats and upregulates pulmonary glucocorticoid receptors. Respirology (Carlton, Vic.), 13(6), 772–780. https://doi.org/10.1111/j.1440-1843.2008.01344.x

Whyte, C. S., Morrow, G. B., Mitchell, J. L., Chowdary, P., & Mutch, N. J. (2020). Fibrinolytic abnormalities in acute respiratory distress syndrome (ARDS) and versatility of thrombolytic drugs to treat COVID-19. Journal of Thrombosis and Haemostasis, 18(7), 1548-1555.

Wiscombe, S. (2021). A Lipopolysaccharide (LPS) inhalation model to characterise divergent innate cellular responses and presence of alveolar leak, early in the course of acute lung inflammation (Doctoral dissertation, Newcastle University).

Zou, B., Jiang, W., Han, H., Li, J., Mao, W., Tang, Z., Yang, Q., Qian, G., Qian, J., Zeng, W., Gu, J., Chu, T., Zhu, N., Zhang, W., Yan, D., He, R., Chu, Y., & Lu, M. (2017). Acyloxyacyl hydrolase promotes the resolution of lipopolysaccharide-induced acute lung injury. PLoS pathogens, 13(6), e1006436. https://doi.org/10.1371/journal.ppat.1006436

# 10

# A Laboratory Protocol for Mesenchymal Stem Cell Isolation from Veterinary Important Species

***Milindmitra K Lonare, Afroz Jahan, Manjinder Sharma, and Sivaraman Ramanarayanan***

## Mesenchymal Stem Cells

Recent advances in the field of regenerative medicine have opened up newer therapeutic modalities using cell-based and cell-free therapies, particularly stem cells, as potential candidates. As per the definition, stem cells are the undifferentiated progenitor cells that can regenerate and can also differentiate into various specialized cells (Hoang *et al.*, 2022). They can be classified as embryonic stem cells (ESCs), induced pluripotent stem cells (iPSCs) and postnatal adult stem cells (Locke *et al.*, 2009). Among postnatal adult stem cells, mesenchymal stem cells (MSCs) have emerged as a popular choice of candidate for therapeutics due to the relatively easy availability, simple isolation techniques, low risk of teratogenicity, minimal ethical and legal concerns unlike embryonic stem cells, hence, are easy to apply in clinical research and practice (Voga *et al.*, 2020). Among MSCs of different origin, bone marrow and particularly adipose tissue-derived stem cells (ADMSCs) received more attention and are one of the most common and promising stem cell types. These cells can be simply obtained from subcutaneous and omental fat or liposuction aspirates for the treatment of diverse clinical applications (Kim &Heo, 2014). These cells can also be expanded vigorously in vitro and differentiated into specific cell lineages such as adipocytes, osteoblasts, chondrocytes, and neurocytes and are even genetically stable in long-term cultures (Miana& González, 2018). Mesenchymal stem cells (MSCs) are multipotent progenitor cells possessing self-renewal ability (limited in vitro) and differentiation potential into mesenchymal lineages, according to the (ISCT) International Society for Cell and Gene Therapy (Dominici *et al.*, 2006).

## Potential Sources

MSCs can be effectively isolated from a wide range of tissues, depending on their origin, they have different characteristics. The source of tissue should be taken into account when choosing the appropriate stem cell therapy for tissue repair. MSCs can be isolated from tissues like bone marrow (Singh *et al.*, 2021), adipose tissue (Sampaio *et al.*, 2015), synovial fluid (Bearden *et al.*, 2017), umbilical cord (Pham *et al.*, 2019), dental pulp (Petchdee&Sompeewong, 2016), Wharton's jelly (Kang *et al.*, 2012) etc. Among these, adipose tissue is one of the common and easiest sources of stem cells that can collected from the animals during routine surgical procedures of different species of animals presented in veterinary hospitals. While, bone marrow collection requires some specialized procedure and technical procedure with expertise as it is usually collected by aspiration of the iliac crest of pelvis crest of the animals.

## Characterization

It is mandatory to confirm the cells isolated are stem cells or other types. To resolve this issue guidelines for characterization of MSCs have been specified by the ISCT, wherein it is recommended to observe for the expression of a) adherence to plastic surfaces, b) expression of a minimum of two or three positive surface markers i.e. CD 73, 105, 90 etc. and negative markers such as CD 34 and 45 and c) differentiation into adipogenic, osteogenic and chondrogenic lineages, characters that differentiate them from cells of Hemopoetic origin (Dominici *et al.*, 2006). The presence of markers such as CD90, CD105, CD44, Nanog, OCT4 in MSC of bovine (Gade *et al.*, 2013), canine (Reich *et al.*, 2012), equine (Radtke *et al.*, 2013) indicates the confirmation of stem cell type. Many researchers reported variations in the expression of stem cell markers in freshly isolated and cultured MSCs.

## Therapeutic Application

Vast literature is available, particularly in the field of veterinary medicine where MSCs have been reported to be useful in mitigating and repairing damage. Now, MSCs are under use for animals and humans healthcare for treating conditions orthopaedic, digestive, hepatic, renal, cardiac, respiratory, neuromuscular, dermal, ligament injuries and reproductive disorders. Many reports related to uses of MSCs for the treatment of different conditions are available like autologous bone marrow and adipose derived stem cells into a horse's superficial digital flexor tendon (Kornicka-Garbowska *et al.*, 2019), single intra-articular injection of allogeneic bone marrow MSCs for dogs tibial plateau levelling osteotomy (Taroni *et al.* 2017), autologous adipose derived MSCs IV administration to dogs in experimentally induced acute hepatic damage (Yan *et al.* 2019), subconjunctival delivery of autologous BM-MSC for improvement of immune-mediated keratitis in horses (Davis *et al.*, 2019), chronic wound healing with MSCs in goats, buffaloes, Sheep and Horses (Pratheesh *et al.*, 2017; Martinello *et al.*, 2018; Lanci *et al.*, 2019) and mild to moderate improvements in gait, nociception, and proprioception in some of the animals (Besalti *et al.*, 2016).

## Protocols

### Minimum Requirement of Instruments and Chemicals

*Instruments*: Biosafety cabinet (BCL-II grade), inverted microscope, $CO_2$ incubator, $CO_2$ cylinders with regulators, cooling centrifuge machine, autoclave, hot air oven, hot plat with magnetic steror, and temperature controller.

*Chemicals*: Milli-Q water, Dulbecco's Modified Eagle's Medium (DMEM)-high glucose, Trypsin EDTA or trypsin versene solution, di-methyl sulfoxide (DMSO), neutral red dye, Trypan blue, gentamicin, L-glutamine, phenolred, sodiumpyruvate, sodium bicarbonate, fetal bovine serum (FBS)

*Glass-plasticware*: Glass beakers (500ml, 100ml, 50ml), screw-capped reagent bottles (250ml, 500ml, 1000ml), funnels, micropipettes, micro-tips (200µl, 1000 µl),centrifugetubes (2ml, 15ml), serological pipettes, cryovials, sterile syringe filters (0.22 µm), tissue culture flasks (25 $cm^2$, 75 $cm^2$ and 96 well culture plates).

### Composition of Complete Growth Medium

The common growth medium for culturing is using Dulbecco's Modified Eagle Medium-high glucose (4500 mg/L) withL-glutamine (584mg/L), phenolred (14.95mg/L), sodium pyruvate (110mg/L), sodium bicarbonate (3700 mg/L) without HEPES.

### Composition of Complete Growth Medium (CGM)

| | |
|---|---|
| Fetal Bovine Serum (FBS) | 15 mL |
| DMEM-H Gaddto | 100 mL |
| Gentamicin (50 mg/mL) | 100 µL |

(if needed: Gentamicin or antimycotic-antibiotic solution can be added in CGM depending upon the need of the laboratory and level of contamination)

The complete growth medium (CGM) after preparation, filtered through a sterile 0.2 µm membrane filter and stored in screw-capped glass reagent bottles of 100 to 250 ml. A small aliquot of the CGM has to be prepared and incubated at 37 °C with 5% $CO_2$ for 48 hrs to check for preliminary bacterial contamination which would be possible by visualizing the changes of the medium such as pH change and visual turbidity of the medium. For routine work, store the CGM at 4 °C and thaw to 37 °C before it is added to the flack.

### Composition of Phosphate-buffered Saline (PBS)

Prepare the sterile PBS solution needed during various stages of MSCs culturing, for cell isolation, cell washing, preparation of other reagents and staining solution etc.

Composition of 1X PBS solution is as follows,

| | |
|---|---|
| Sodium chloride (NaCl) | 8.00g |
| Potassium chloride (KCl) | 0.20g |
| Potassium dihydrogen phosphate anhydrous (KH2PO4) | 0.24g |
| Disodium Hydrogen phosphate anhydrous (Na2HPO4) | 1.44g |

Dissolve theabove-mentionedsaltsinitiallyin800mLofMilli-Q water, and pH is adjusted to 7.4 using 0.1N HCl or NaOH solution, and then the final volume should be made up to 1000 mL with Milli-Q water and 0.5% phenol red solution in Dulbecco's PBS @ 1 μL/mL to monitor the pH changes during storage. Finally, autoclave PBS at 120 OC, 15 lbs for 20 min. Before use filtered with a 0.2 μm membrane siring filter, store in refrigerator. Before use, bring the temperature of the PBS at room temperature. If needed, gentamicin or antimycotic-antibiotic solution with pH indicator (phenol red) can be added in PBS depending upon the need of the laboratory and the level of contamination.

### Composition of RBC Lysis Buffer

RBC lysis buffer (100 ml) can be prepared as below

| | |
|---|---|
| Ammonium chloride ($NH_4Cl$) | 0.826 gm |
| Sodium bicarbonate ($NaHCO_3$) | 0.1 gm |
| EDTA | 0.004 gm |

Mix all the salts one by one in 80ml triple distilled water and adjust the pH to 7.3 with the help of NaOH and HCl. Make the final volume upto 100 ml. Sterilize by autoclaving, cool, and store at room temperature for several days.

### Enzymes for tissue digestion

Enzymes like trypsin and Collagenase are essential for solid tissue digestion during cell isolation and detachment of adherent cells during subculture. Recombinant trypsin, solutions containing EDTA, and non-enzymatic solutions provide stabler, gentler options for wider culture applications. Collagenase cleaves the peptide bonds in native, triple-helical collagen. Because of its unique ability to hydrolyze native collagen, it is widely used in isolation of cells from animal tissue. Dispases are rapid, effective, gentle, and neutral proteases that can separate intact epithelial sheets in culture from the substratum. The most potent Collagenase is the "crude" Collagenase secreted by the anaerobic bacterium *Clostridium histolyticum. C. histolyticum* Collagenase have molecular weights from 68,000 to 125,000 Da and are metalloproteinases that require zinc and calcium. Collagenase supports a wide range of research as it can be used to isolate cells/tissues by degrading collagen in the skin, blood vessels, tendons, and bones. The type of Collagenase is chosen based on the specific type of cell or tissue that needs to be isolated. Collagenase type I is suitable for the isolation of fat, lung, epithelial, and adrenal tissue cells. Collagenase type II is ideal for the separation of the liver, thyroid, bone, heart, and salivary gland tissues. Collagenase type IV can break down a variety of tissues, and type V is applicable in the isolation of pancreatic islet tissues by separating the connective tissues into single cells.

## Isolation of MSCs from Bone Marrow and Cord Blood

### Collection and Transportation

Bone marrow samples can be collected from animals presented at the hospital for treatment (autologous) or slaughterhouse animals (allogenic) may be the easiest source for bone marrow cells. Cord blood may be collected at the time of parturition from the naval cord of animals. Shave the part of iliac crest (wing of ileum) with the help of a blade and cleaned the area properly before the surgical procedure. Aspirate the bone marrow cells from iliac crest of pelvis from an animal (buffalo, cattle, dogs or horses etc. of either sex immediately after regional anaesthesia. Use the autoclave surgical cloth to cover the area and use the aspiration needle/Jamshidi needle (11 G) with 20-ml syringes containing 250 units of heparin. Pierce the Jamshidi needle with a stylet through the skin and tissues on the posterior iliac crest. Once the needle contacts the bone, it is advanced to rotate clockwise and counter-clockwise slowly until the needle enters into bone marrow cavity. Usually, a sudden loss of resistance or giveaway when the needle enters the bone marrow. Aspirate the 10 ml of bone marrow into a syringe containing anticoagulant (Heparin) under aseptic conditions. After the collection of the bone marrow sample mix properly with the anticoagulant to avoid coagulation of sample and transport to the laboratory as early as possible to minimize the viability (Fig. 1).

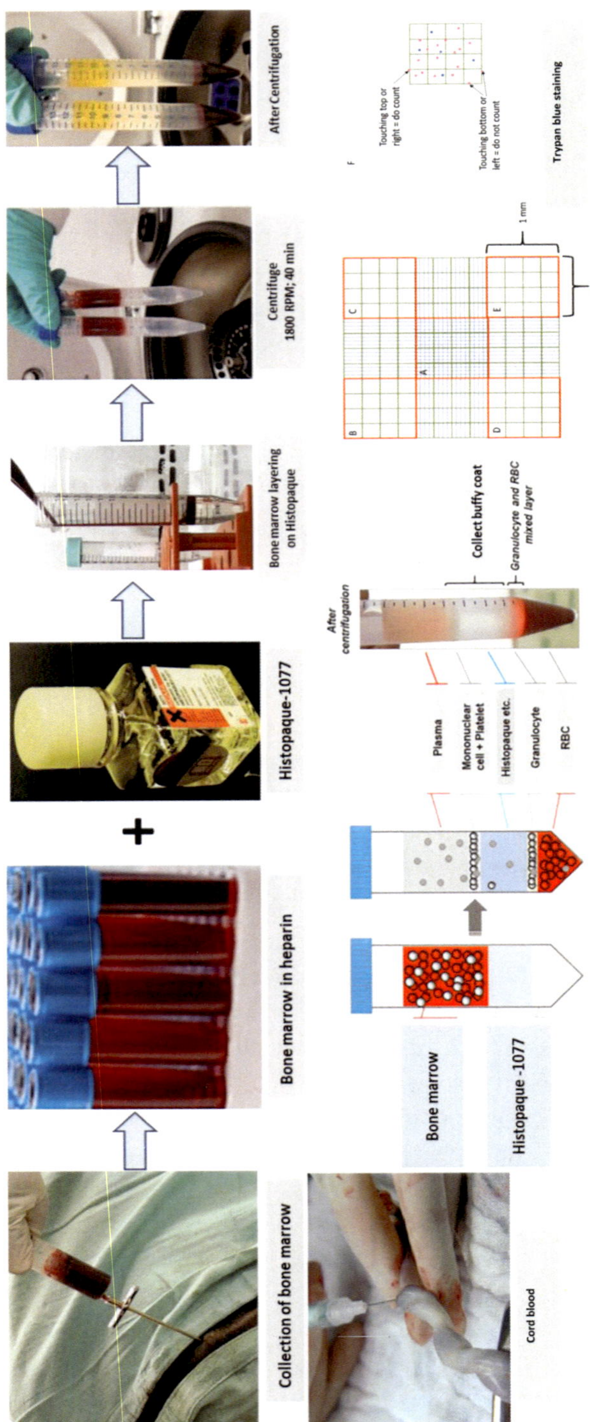


**Figure 1:** Schematic representation of bone marrow and cord blood collection and MSCs isolation from the bone marrow.

## Isolation and Culturing

Collect bone marrow samples and transport into the laboratory in aseptic carrier case. Bone marrow samples consist of a mixed population of cells. Isolate the MSCs by density gradient method using Histopaque-1077. Take 5 ml Histopaque-1077 reagent in 15 centrifuge tubes and allow it to stand on test-tube stand to attend room temperature. Layer the 5ml bone marrow sample on Histopaque-1077 (1:1 ratio) slowly. Centrifuge the sample (15ml test tube) @ 1800 rpm for 30-40 minutes at room temperature. Carefully remove test-tube without any disturbance. After centrifugation, the mononuclear cell suspension is visible as a buffy coat at the interphase of two solutions consisting of a mixed population of cells (lymphocytes and stem cells). Transfer the buffy coat carefully to another 15 ml centrifuge vial and, wash with PBS (1:3 ratio) by gentle pipetting. Centrifuge @ 900 rpm for 10 minutes at room temperature. Resuspend the cell pellet in red blood cell (RBC) lysis buffer. Incubate at room temperature for 5 min and centrifuge the tube @800 rpm for 10 min. Thereafter wash the cell to clear off RBC lysis buffer by washing with PBS twice following resuspension in DMEM for final washing. Finally, isolated cells resuspend in 1ml CGM (Fig. 1). Cell viability will be done by trypan blue dye exclusion assay method. If the viable cell count is more than 1 million then use for seeding. Keep flask un-disturbed for 48-72 hrs. Remove the floating dead cells by gentle washing with PBS. Thereafter, follow the partial removal of CGM and the addition of fresh CGM every 3$^{rd}$ day until the cell grows into a full monolayer (80-90% confluency). After this, trypsinize the cell using 0.25% trypsin-EDTA and sub-culture the cells.

## Isolation of Mesenchymal Stem Cells from Adipose Tissue

### *Collection and Transport of Adipose Tissue Samples*

The adipose tissue samples, irrespective of sex and breed of animals can be collected directly from slaughterhouses for large animals or animals presented at hospitals for surgical intervention or operation. Usually, a lot of facia and adipose tissues are dissected during surgery as a surgical iste. Tissue weight of approximately 10-15g from large animals and 5-6g from small animal's adipose tissue samples can be collected in the sterile PBS with antimycotic-antibiotic solution (pH7.4,at37°C). The collected subcutaneous layer or omentum fat sample in sterile sample containers should be transported to the lab in a sterile safety carrier case. Then wash the collected samples by serial washing using PBS-antibiotic solution inside the biosafety cabinet. Cut the tissue into small pieces as much as possible. Wash samples thrice again to minimize the contamination and excess of facia. Finally, transfer the issues to 50 ml conical flask having enzyme solution for tissue digestion (Fig. 2).

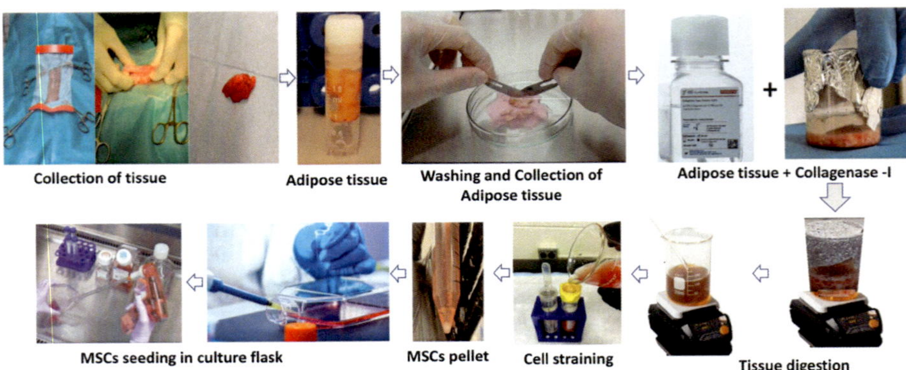


**Figure 2:** Schematic representation adipose tissue collection and MSCs isolation in different species of animal.

## Enzymatic Digestion of Buffalo Tissue

Process the adipose tissue of buffalo origin by the method of enzymatic digestion as described earlier (Deng *et al.*, 2018 and Sampaio *et al.*, 2015). The method may be optimized depending on the laboratory facilities (Fig. 3). Wherein, approximately 5-10 grams of tissue is taken and washed several times with sterile PBS-antibiotic solution, pH7.4, to remove excess blood and unwanted tissues. Finely chopped the tissues into small pieces (3-5 mm size) using sterile forceps and scissors to facilitate uniform enzymatic digestion. Add the tissues in 50ml conical flask with 25 ml 0.1% Collagenase type I enzyme solution. Again, wash the tissue with sterile PBS and then transfer it into a sterile glass beaker having magnetic beads. Close the mouth of the beaker with sterile aluminum foil and sealed with parafilm. Place the digestion mixture in a CO2 incubatorat37°Cfor10minutes. Then place beaker on a hot plate with a magnetic stirrer (37 $^{0}$C) for 5 min for stirring. The digestive enzyme-tissue mixture should be regulated under 37 $^{0}$C inside the $CO_2$ incubator and outside the incubator for optimum cell yield. Temperature variation may affect the optimum enzymatic activity and tissue digestion. Repeat the cycle of incubation and hot plate steering until the tissue looks completely digested. Complete digestion indicated by clear or straw-colored solution on visual inspection with formation of two phases, a floating froth phase consisting of digested adipose tissue and a lower second phase (straw color) containing stromal vascular fraction (SVF) that consist of MSCs, erythrocytes, leukocytes,and other cellular fractions. Finally, add an equal volume of CGM to neutralize the collagenase-I enzyme activity to prevent further cell damage. Digested sample filter with sterile cell strainer into a conical flask. Transfer the resultant filtrate into 15 ml centrifuge tubes. Centrifuge the vials @ 1000 rpm for 10 min. Discard the supernatant and resuspend the cell pellet in 4 ml red blood cell (RBC) lysis buffer. Incubate the cells in an incubator for 5 min. Again, gently pipette to lyse the RBC. Centrifuge the mixture @ 1000 rpm for 10 min. Decant the supernatant and washed the resultant cell pellet in 2-3mlsterilePBS (pH7.4) and centrifuge @ 800 rpm for10 minutes. Repeat the last step at least 3 times to remove the excess of RBC lysis solution and impurities. Last washing should be given with 1 ml DMEM. Finally, resuspend the pellet in 1m CGM. Check the viability with trypan blue dye exclusion technique (Strober, 2015). The resultant cell suspension transfer to 25 cm$^2$ tissue culture flaskswith the addition of 4 ml CGM and incubated @ 37 °C with 5 % CO2. Keep flask undisturbed and allow to attach to the surface for the next 48-72 hrs (Fig. 2). Remove the initial media to remove the non-adherent cells and cell debris, if any. Thereafter the halfway medium should change whenever necessary until the cell stained the required confluency (>80%).

## Enzymatic Digestion of Dog and Poultry Tissue

Dog and poultry adipose tissue is softer than buffalo. Collect the adipose tissue and follow a similar procedure for serial washing as discussed in an earlier section. Then chapped the tissue into smaller pieces of 4-5 mm size and again washed it. Place all cleaned tissue in a 50 ml conical flask having 25 ml of 0.1 collagenase type I solution. Keep the digestion mixture in the incubator for 5 minutes. Then transfer on a hot plate with having magnetic stirrer. Regulate the optimum temperature of the hot plate to 37 °C. Agitate the sample with a magnetic beat continuously stirring for the next 20 min. Thereafter tissue will be completely digested and can be confirmed by complete homogenous fluid with no different phases. Straw color phase containing stromal vascular fraction (SVF) that consists of MSCs, erythrocytes, leukocytes, and other cellular fractions. Stopped the enzymatic activity by adding the excess of pure 1ml FBS or an equal volume of CGM that prevented excessive cell damage. Filter the complete fluid with a sterile cell strainer or a thin layer of cotton mesh into a conical flask. Then transfer the resultant filtrate into 15 ml centrifuge tubes and centrifuge @ 1000 rpm for 10 min. Discard the supernatant and resuspend the cell pellet in4 ml red blood cell (RBC) lysis buffer. Incubate the cells in an incubator (dog sample 2-3 min, poultry sample 5-6 min). Again, gently pipette to lyse the RBC completely. Centrifuge the mixture @800 rpm for 10 min. Decant supernatant and wash the resultant cell pellet with 2-3mlsterilePBS (pH7.4) and centrifuge @ 800 rpm for 10 min.

Repeat the last step at least 3 times to remove the excess of RBC lysis solution and other impurities. Last washing should be given with 1 ml DMEM. Finally, resuspend the pellet in 1m CGM. Check the viability with trypan blue dye exclusion technique (Strober, 2015). The resultant cell suspension transferred to 25 $cm^2$ tissue culture flaskswith the addition of 4 ml CGM and incubated @ 37 °C with 5 % CO2. Keep the flask undisturbed and allow it to attach to the surface for the next 24-48 hrs (Fig. 2). Remove completely all the media to clear the non-adherent cells and cell debris, if any. If, any mild contamination, then wash with PBS 2-3 times. Thereafter the halfway medium should change whenever necessary until the flask attains the required confluency (>80%).

### Sub-culturing of MSCs

The cells can be sub-culture after the yatta in 70-80% confluency. The expended medium can be removed and the cells may be washed with sterile PBS. Add the 2ml of trypsin phosphate versine glucose (TPVG) solution (0.25% trypsin, 0.5% PVP,0.02% EDTA, and 0.05% Glucose in DPBS) to the25$cm^2$tissuecultureflasks or 4 ml TPVG solution for 75 $cm^2$tissue culture flasks and incubate in the $CO_2$ incubator for optimum activity and checked for 3-65min (dog < cattle; depending on the origin of samples) to facilitated is sociation of the cells. After visualizing the cell dissociation under aphase contrast in verted microscope (Nikon Eclipse Ti, Japan) in the form of rounding of cells and free movement. To neutralize the TPVG enzymatic activity, add an equal volume of sterile CGM or a small quantity of pure FBS. Pipette the cells in a flask for complete removal of adherent cells and cell suspension transfer to a 15 ml centrifuge tube. Then, centrifuge @ 800rpm for10 minutes. Discard the supernatant and resuspend the cell pellet in 1 ml PBS. Repeat this step thrice and finally resuspend the cells in CGM and count the viable cells by hemocytometer by the trypan blue dye exclusion assay. Wash the flask thrice with PBS to remove the excess trypsin in the culture flask. The seeding density should be maintained at 1.5 x$10^6$ viable cells per 25 $cm^2$tissue culture flask.

### Cell Viability Assay

The cell viability is assessed by the trypan blue dye exclusion method as described by Strober (2015).

### Principle

In this method, cell viability is determined indirectly by assessing cell membrane integrity. Trypan blue is a diazo dye that is widely used to selectively stain non-viable cells. The dye is negatively charged and non-permeable unless the integrity of the cell membrane is compromised. The viable cells have intact cell membranes that exclude trypan blue and appear unstained, while the non-viable cells appear blue in colour.

## Procedure

Take 20μL aliquot from 1mL of a uniform cell suspension and mixed with an equal volume of 0.1% trypan blue solution. The trypan blue-cell suspension mixture then charged onto the Neubauer counting chamber of the hemocy to meter. Placed the hemocytometer on the stage of a binocular compound microscope and count the cells in four primary squares (WBC counting squares). Count the viable (unstained) and non-viable (blue-stained) cells separately (Fig. 3). Calculate the total number of cells and percent of viable cells per mL using the formula mentioned below.

Total number of cells=Average of cells counted in four primary squares X D ilution factor X$10^4$

Whereas,

$$\text{Percentage of viable cells (per mL)} = \frac{\text{Total number of viable cells per mL X100}}{\text{Total number of cells per mL}}$$

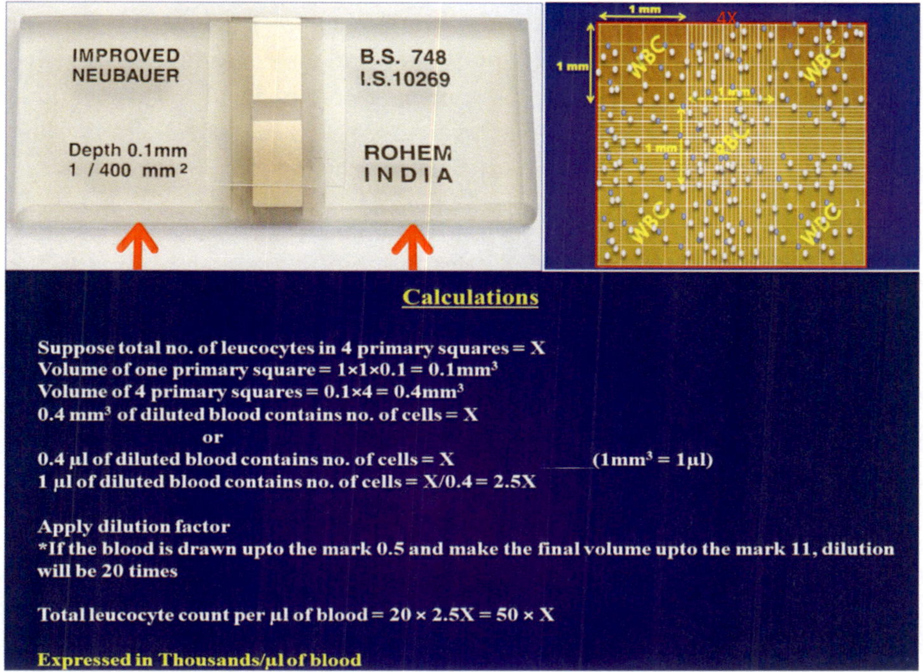


**Figure 3:** Trypan blue dye exclusion showing the live (transparent) and dead (blue) cells in hemacytometer with counting chambers.

## Cell Morphology and Growth Pattern

### *Morphometry*

MSCs morphology can analyzed after initial isolation and after every passage by observing under inverted phase contrast microscopy. The growth pattern of cells can be studied wherein the number of days in between passages and observation can be recorded. The morphometry of cells (cell sizing) can also studied using NIS-Elements software (Nikon,Japan) wherein, cells at initial isolation and passages with trypsinization for confirmation or suitability for administration to the patients.

## Growth Kinetics Assay

Growth kinetics is also known as population doubling time (PDT).The time taken for the cell population to become double in the count and it is estimated by calculating the concentration of viable cells by colorimetric method. The population doubling time of MSCs can be calculated using a available commercial CCK-8kit or MTT assay.

The principle behind the assay is that the WST-8, amonosodium tetrazolium salt is bio-reduced by cellular dehydrogenases to form anorange-colouredfor mazan product that is soluble in tissue culture medium.The intensity of color produced by the reaction is directly proportional to the concentration of viable cells present. This assay is advantageous over conventional MTT-assay due to relatively less cytotoxicity and since the formazan produced here is water soluble there is no requirement of a specialized solvent to solubilize.

## Procedure

The PDT of MSCs can be studied after the third passage according to the method of Washiyama *et al.* (1996) with modifications by our laboratory. Briefly, seed the 4 rows of 96wellplate with a known number of viable cells (6000 cells/well) and after 6 h of seeding, when the cells get attached to the surface, the initial 2 rows used for initial reading and next 2 rows wells keep unused for final cell population reading after48hrs. Remove the old CGM and add new CGM and then add 10µLCCK-8ready to used solution to each of the well in first 2 rows. Then plate incubate for 4 hrs @ 37 °C and 5% CO2 in a humidified atmosphere as per instruction. The read the OD measuring the OD of the wells at 450 nm in a multiplate reader (Multiskan Go, Thermo Fwasher Scientific, USA). Repeat the same step in remaining 2 rows after completion of 48 hrs. Remove the old media and add new media and add 10µLCCK-8readytousedsolutiontoeachofthewell in 3 and 42 rows. Incubate the plate for next 4hrs and measure the OD. Calculate initial and final cell population. The harvested final cell number can be calculated using following formula.

$$\text{Harvested cell number} = \frac{\text{Final OD X Initial cell number}}{\text{Initial OD}}$$

From there, the population doubling time can be calculated using the equation(Cristofalo *et al.*, 1998)

$$\text{Population Doubling Time} = \frac{\text{Culture time (CT)}}{\text{Cell doubling (CD)}}$$

Wherein, CD = [log (NH/NI)] ÷ 2; NH is the harvested cell number, and NI is the initial cell number.

## MMT Assay

### Principle

MMT Assay is also known as Cell Viability or Cytotoxicity Assay. Rapid colorimetric assay based on the cleavage of the tetrazolium ring of MTT (3-(4,5-dimethylthazolk-2-yl)-2,5-diphenyl tetrazolium bromide) by dehydrogenases in active mitochondria of living cells as an estimate of viable cell number. It is an ability of nicotinamide adenine dinucleotide phosphate (NADPH)-dependent cellular oxidoreductase enzymes to reduce the tetrazolium dye MTT to its insoluble formazan, which has a purple color.

### Procedure

For the determination of the total viable cells in a culture this assay is used routinely. Seed the MSCs at a concentration of $1 \times 10^6$ cells/ml in culture medium in 96 well culture plate at least for 6-12 hrs and then treat the cells with target agent with different concentrations (eg. xenobiotic concentrations 2.5, 5, 10, 25, 50, 100, 200 µg/mL). The time of exposure will be different depending on the protocols. Incubate MSCs for the required time at 37 °C and 5% $CO_2$. After the treatment period is over, remove the media and 100µL DMEM (4 % FBS). Then, add 20µL of 5 mg $mL^{-1}$ of MTT solution prepared in DMEM (4 % FBS) without phenol red and incubate the plate by covering it with aluminum foil at 37 $^0$C in 5% $CO_2$ humidified chamber/incubator. After the incubation period of 3 hrs, the MTT solution is carefully removed. Formazan crystal formed intracellularly has to be dissolved by adding a solubilizing agent (200 µL of DMSO), cover a plate with aluminum foil and incubate for 20-30 min in a dark inside incubator until complete dissolution is achieved. Finally, read the absorbance at 570 nm (test wavelength) and 720 nm (reference wavelength) using Tecan Infinite M200 Multiplate Reader (Fig. 4). The absorbance values and all results can be presented as percentage of the control values (viability/ control; % of cell survival). The percentage of cell survival can be expressed with the following formula and $IC_{50}$ of the xenobiotics can be calculated using the following formula.

$$\text{Percentage of cell survival} = \frac{\text{Absorbance}(570\,\text{nm})\,\text{tested compound X100}}{\text{Absorbance}(570\,\text{nm})\,\text{control}}$$

$IC_{50}$= 100 - % of cell survival

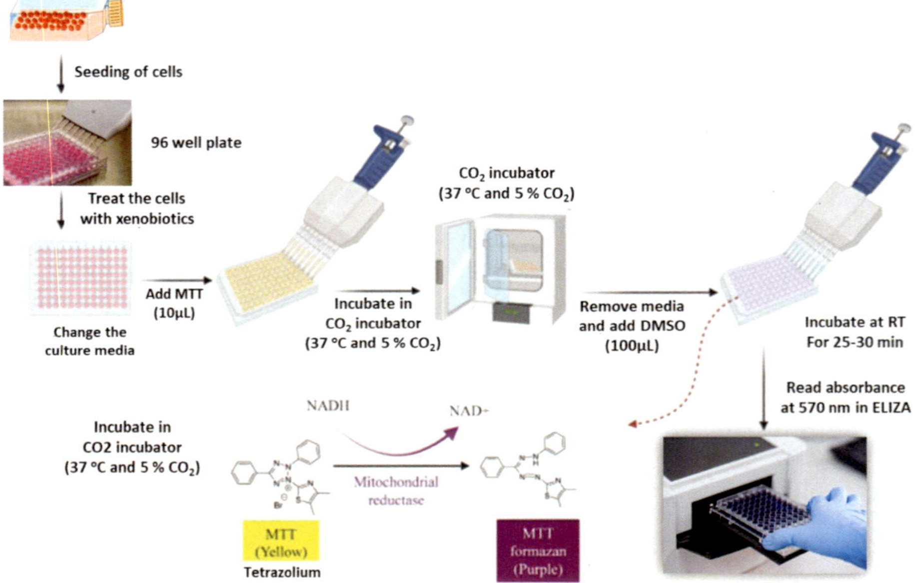


**Figure 4:** MTT assay protocol for determining cell viability and median inhibitory concentration.

## References


Bearden, R. N., Huggins, S. S., Cummings, K. J., Smith, R., Gregory, C. A., & Saunders, W. B. (2017). In-vitro characterization of canine multipotent stromal cell sisolated from synovium, bone marrow, and adipose tissue :adonor-matched comparative study. Stem Cell Research & Therapy, 8,218. https://doi.org/10.1186/s13287-017-0639-6

Besalti, O., Aktas, Z., Can, P., Akpinar, E., Elcin, A.E., & Elcin,Y.M.(2016). The use of auto logous neurogenically-induced bone marrow-derived mesenchymal stem cells for the treatment of paraplegic dogs without nociception due to spinal trauma. Journal of Veterinary Medical Science, 78(9), 1465–1473.https://doi.org/10.1292/jvms.15-0571

Davis, A. B., Schnabel, L. V, & Gilger, B. C. (2019). Sub conjunctival bone marrow -derived mesenchymal stem cell therapy as an ovel treat mental ternative for equineimmune-mediated keratit is : Acaseseries. Veterinary Opthalmology, 22(5), 674–682. https://doi.org/10.1111/vop.12641

Deng, Y., Huang, G., Zou, L., Nong, T., Yang, X., Cui, J., Wei, Y., Yang, S., & Shi, D. (2018). Isolation and characterization of buffalo (Bubalusbubalis)amniotic mesenchymal stem cells derived from amnion from the first trimester pregnancy. Journal of Veterinary Medical Science, 80(4), 710–719. https://doi.org/10.1292/jvms.17-0556

Dominici, M., LeBlanc, K., Mueller, I., Slaper-Cortenbach, I., Marini, F.C., Krause, S., Deans, R.J., Keating, A., Prockop, D.J., & Horwitz, E.M. (2006). Minimal criteria for defining multipotent mesenchymal stromal cells. The International Society for Cellular Therapy position statement. Cytotherapy, 8(4), 315–317. https://doi.org/10.1080/14653240600855905

Gade, N.E., Pratheesh, M.D., Nath, A., Dubey, P.K., Amarpal, Sharma, B., Saikumar, G., &Taru Sharma, G. (2013). Molecular and Cellular Characterization of Buffalo Bone Marrow-Derived Mesenchymal Stem Cells.Reproduction in Domestic Animals, 48, 358–367. https://doi.org/10.1111/j.1439-0531.2012.02156.x

Hoang, D. M., Pham, P.T., Bach, T.Q., Ngo, A.T.L., Nguyen, Q. T., Phan, T. T. K., Nguyen, G.H., Le, P.T.T., Hoang, V.T., Forsyth, N.R., Heke, M., & Nguyen, L. T. (2022). Stem cell-based therapy for human diseases. Signal Transduction and Targeted Therapy, 7(272). https://doi.org/10.1038/s41392-022-01134-4

Kang, K.S., & Trosko, J.E. (2011). Stem cells in toxicology:Fundamental biology and practical considerations. Toxicological Sciences, 120(S1), 269–289. https://doi.org/10.1093/toxsci/kfq370

Kornicka-Garbowska, K., Pędziwiatr, R., Woźniak, P., Kucharczyk, K., & Marycz, K. (2019). Microvesiclesisolated from 5- azacytidine- and- resveratrol- treated mesenchymal stem cells for the treatment of suspens or yligament injury in horse-Acasereport. Stem Cell Research and Therapy, 10(394). https://doi.org/10.1186/s13287-019-1469-5

Lanci, A., Merlo, B., Mariella, J., Castagnetti, C., &Iacono, E. (2019). Heterologous Wharton' s Jelly Derived Mesenchymal Stem Cells Application on a Large Chronic Skin Wound in a 6-Month-Old Filly. Frontiers in Veterinary Science, 6,9. https://doi.org/10.3389/fvets.2019.00009

Locke, M., Windsor, J., & Dunbar, P. R. (2009). Human adipose-derived stem cells: Isolation, characterization and applications in surgery. ANZ Journal of Surgery, 79(4),235–244. https://doi.org/10.1111/j.1445-2197.2009.04852.x

Martinello, T., Gomiero, C., Perazzi, A., Iacopetti, I., Gemignani, F., Debenedictis, G.M., Ferro, S., Zuin, M., Martines, E., Brun, P., Maccatrozzo, L., Chiers, K., Spaas, J.H., & Patruno, M. (2018). Allogeneic mesenchymal stem cells improve the wound healing process of sheep skin. BMC Veterinary Research, 14,202.

Miana, V.V., & González, E.A.P. (2018). Adipose tissue stem cells in regenerative medicine. Ecancermedical science, 12,822. https://doi.org/10.3332/ecancer. 2018.822

Petchdee, S., & Sompeewong, S. (2016). Intravenous administration of puppy deciduous teeth stem cells in degenerative valve disease. Veterinary World, 9(12), 1429–1434. https://doi.org/10.14202/vetworld.2016.1429-1434

Pham, L.H., Vu, N.B., & Pham, P.Van. (2019). The sub population of CD105 negative mesenchymal stem cells show strong immunomo dulation capacity compared to CD105 positive mesenchymal stemcells. Biomedical Research and Therapy,6(4),3131–3140.

Pratheesh, M.D., Dubey, P.K., Gade, N.E., Nath, A., Sivanarayanan, T.B., Madhu, D. N., Somal, A., Baiju, I., Sreekumar, T. R., Gleeja, V. L., Bhatt, I. A., Chandra,V., Amarpal, Sharma, B., Saikumar, G., & Sharma, G. T. (2017). Comparativestudy on characterization and wound healing potential of goat ( Capra hircus )mesenchymal stem cells derived from fetal origin amniotic fluid and adult bonemarrow. Research in Veterinary Science, 112, 81–88. https://doi.org/10.1016/j.rvsc.2016.12.009

Radtke, C.L., Nino Fong, R., Esparza Gonzalez, B.P., Stryhn, H., & Mcduffee, L.A. (2013). Characterization and osteogenic potential of equinemuscle tissue- and periosteal tissue-derived mesenchymal stem cells in comparison with bonemarrow-andadiposetissuederived mesenchymal stem cells. American Journal of Veterinary Research, 74(5), 790–800. https://doi.org/10.2460/ajvr.74.5.790

Reich, C.M., Raabe, O., Wenisch, S., Bridger, P.S., Kramer, M., & Arnhold, S. (2012). Isolation, culture and chondrogenic differentiation of canine adipose tissue- andbone marrow-derived mesenchymal stem cells-A comparative study. Veterinary Research Communications, 36(2), 139–148.https://doi.org/10.1007/s11259-012-9523-0

Sampaio, R.V., Chiaratti, M.R., Santos, D.C.N., Bressan, F.F., Sangalli, J.R.,Sá, A.L.A., Silva, T. V. G., Costa, N. N., Cordeiro, M. S., Santos, S. S. D., Ambrosio, C.E., Adona, P.R., Meirelles, F.V., Miranda, M.S., & Ohashi, O. M. (2015). Generation of bovine (Bosindicus) and buffalo (Bubalusbubalis) adipose tissue derived stem cells: Isolation, characterization, and multipotentiality. Genetics and Molecular Research, 14(1),53–62. https://doi.org/10.4238/2015. January.15.7

Sampaio, R.V., Chiaratti, M.R., Santos, D.C.N., Bressan, F.F., Sangalli, J.R., Sá, A.L. A., Silva, T. V. G., Costa, N. N., Cordeiro, M. S., Santos, S. S. D., Ambrosio, C.E., Adona, P.R., Meirelles, F.V., Miranda, M.S., & Ohashi, O. M. (2015). Generation of bovine (Bosindicus) and buffalo (Bubalusbubalis) adipose tissue derived stem cells: Isolation, characterization, and multipotentiality.Genetics and Molecular Research, 14(1), 53–62. https://doi.org/10.4238/2015. January.15.7

Singh H, Lonare MK, Sharma M, et al. Interactive effect of carbendazim and Imidacloprid on buffalo bone marrow derived mesenchymal stem cells: oxidative stress, cytotoxicity and genotoxicity. Drug ChemToxicol. 2021;2021:1-15. doi:10.1080/01480545.2021.2007023.

Strober, W. (2015). Trypan Blue Exclusion Test of Cell Viability. Current Protocols in Immunology, 111(1), A3.B.1-A3.B.3. https://doi.org/10.1002/0471142735.ima03bs111

Taroni, M., Cabon, Q., Fèbre, M., Cachon, T., Saulnier, N., Carozzo, C., Maddens, S., Labadie, F., Robert, C., &Viguier, E. (2017). Evaluation of the effect of a singleintra-articular injection of all ogeneicneonatal mesenchymal stromal cells compared to oral non-steroidal anti-inflammatory treatment on the postoperative musculoskel etal status and gait of dogs over a 6-month period a. Frontiers in Veterinary Science, 4, 83. https://doi.org/10.3389/fvets.2017.00083

Voga, M., Adamic, N., Vengust, M., &Majdic, G. (2020). Stem Cells in Veterinary Medicine Current State and Treatment Options. Front. Vet. Sci., 7(May), 278. https://doi.org/10.3389/fvets.2020.00278

Yan, Y., Fang, J., Wen, X., Teng, X., Li, B., Zhou, Z., Peng, S., Arwasha, A.H., Liu, W., & Hua, J.(2019). The rapeutic applications of adipose-derived mesenchymal stem cells on acute liver injury in canines. Research in Veterinary Science, 126, 233–239. https://doi.org/10.1016/j.rvsc.2019.09.004

# 11

# Detection of Environmental Pollutants and Agrochemicals Using Gas Chromatography and Mass Spectrophotometry

*Amit Shukla, Soumen Choudhury, Atul Prakash, Sakshi Tiwari*

Injudicious and indiscriminate use of several agrochemicals *viz.* pesticides, herbicides, insecticides, dyes and medicaments in the field of agriculture and veterinary practices threatens the human and animal population for the toxicity of these environmental deterrents. These agrochemicals have been associated with several episodes of accidental poisoning and mortality among human and animal population. The episode of Mass ethion poisoning in Gujarat is one of the commonest example of inclusion of pesticides or agrochemicals in the environmental pollutants. There are modern tools to evaluate and analyze these chemicals in the environmental pollutants of human and animal use. The sensitivity of these modern highly sophisticated tools has now immensely changed the analytical environmental pollutants industry. Gas chromatography and Mass Spectrophotometry i.e. (GC-MS) technique is employed to detect the volatile residues from environmental pollutants. Whereas, High performance liquid chromatography i.e. HPLC is employed to analyse and detect non volatile drug residues. Gas Chromatography/Mass Spectrometry (GC/MS) instrument separates chemical mixtures (the GC component) and identifies the components at a molecular level (the MS component). It is one of the most accurate tools for analyzing environmental samples.

## Principle of GC-MS

GC-MS technique is a well known analytical tool to determine the volatile substances from the group of similar type of entities. It is a hyphenated and sophisticated tool that amalgamates the GC separation with the highly sensitive detector in the form of mass spectrometer. The samples injected gets ionized and accelerated by the quadrupole assembly and thus gaining different trajectories owing to their mass to charge ratios until they settle down.

GC works on the principle that a mixture of samples will separate into individual substances when heated at high temperature in oven. The heated gases are carried through a column with an inert gas (such as helium) and/ or hydrogen, zero air and nitrogen. As the separated substances emerge from the column opening, they flow into the MS. Mass spectrometry identifies compounds by the mass of the analyte molecule. A NIST library of known mass spectra on the basis of charge mass ratio, covering several thousand of compounds, is stored on a system. Mass spectrometry is considered the only definitive analytical detector.

## Preparation of Samples for GC-MS

Chemicals and Reagents: Chloroform, Methanol, hexane, pyridine, potassium chloride, N-methyl-N- (trimethylsilyl) trifluoroacetamide, lactic acid, methoxamine hydrochloride, valine, butyric acid, urea, glycine, serine, succinic acid, fumaric acid, malic acid, proline, alanine, glutamine, phosphoric acid, fructose, glucose, galactose, gluconic acid, palmitic acid, creatinine, inositol and stearic acid are required for both qualitative and quantitative analysis of volatile substance in milk samples.

## Extraction and Derivatization of Environmental Pollutants Samples for GCMS Analysis

- Samples should be obtained for proper analysis of milk samples.
- To obtain rupture of the milk micelles, 15 ml of sample is sonicated for 15 min; 100 μL of milk is transferred to an Eppendorf tube and then 250 μL of methanol and 125 μL of chloroform should be added.
- Samples should be vortexed every 15 min 4 times and then 380 μL of chloroform and 90 μL of aqueous 0.2 M potassium chloride should be added.
- Centrifuge the sample at 13,572 × g for 10 min at 4°C. After centrifugation, the aqueous layer should be transferred to a glass vial and dried by a gentle nitrogen stream and derivatized with 50 μL of pyridine containing methoxamine hydrochloride at 10 mg/mL. After 17 h, 100 μL of N-methyl-N- (trimethylsilyl) trifluoroacetamide is added and after 1 h, samples are resuspended with 600 μL of hexane.
- This derivatized sample should be used for injection in GC MS for analysis.

## GC-MS Analysis

One μL of derivatized samples should be injected into GCMS instrument in split less mode. The injector temperature should be 200°C. The gas flow rate through the column needs to be adjusted to 1 ml/min. The fused silica capillary column of 0.25-μm DB5-MS column and 30 m × 0.25 mm dimension should be used.

## GCMS Conditioning

-3 min of isothermal heating at 50°C, then increased to 250 at 3°C/min and held at 250°C for 25 min. The transfer line and the ion source temperatures should be 280 and 180°C, respectively. Ions generated at 70 eV with electron ionization and should be recorded at 1.6 scans/s over the mass range m/z 50 to 550. The conditioning may vary lab to lab and standardization is needed for better results optimization.

## Data Analysis and Interpretation

The chromatogram is generated and having the retention time for each peak that can be matched with the NIST library of GC-MS. And according to the retention match type and mass spectrum the hits of the detected compound have been generated and plotted.

## Applications

GC-MS has multiple role in analytical pharmacology to better the identification of the compounds involve in different pharmacotherapeutics and toxicological episodes. The major areas covered by GCMS are as follows:

- Phytoconstituent Analysis of herbs or Polyherbal formulations
- Qualitative screening of drugs and their volatile metabolites in biological fluids
- Residue analysis in milk samples
- Determination of Environmental pollutants
- Forensic investigations
- Screening of vegetables, fruits and grains for different phytoconsitutents

# 12

# Basics of Flow Cytometric Analysis

*Amit Shukla, Mukul Anand, Soumen Choudhury Shalini Vaswani and Sakshi Tiwari*

Flow Cytometry is a tool used to detect and measure chemicophysical characteristics of a population of cellular particles. Modern day biomedical research needs the molecular assembly of tools to proceed in the advent of biomedical research with real time diagnostic, prognostic and experimental detection of markers. For the same, a naïve tool is flow cytometry. In this technique, a sample having cells or particles is suspended into a fluid and injected into instrument. The ability to perform these measurements in a very rapid time span is one of the key advantages of the flow cytometry process. A single-cell suspension must first be prepared in order to analyse solid Tissues. A flow cytometer is look alike to microscope, except that, instead of producing an image of the cell, it generally offers a cell-by-cell basis high-throughput, automated quantification of specified optical parameters.

They can quantify up to three to six properties or components are quantitated in a single sample, cell by cell, for about 10,000 cells, in less than one minute. Sample is focused to ideally flow one cell at a time through a laser beam, where the light scattered is characteristic to the cells and their components. These cells are tagged with fluorescent markers in order to absorb the light and then emitted them in the form of band of wavelengths. Average ten thousand cells can be quickly examined and the data gathered are processed by a computer.

## Components of Flow Cytometer

Flow cytometer has five main components namely:

1. Flow cell
2. Measuring system
3. Detector
4. Amplification system, and
5. Computer for analysis of the signals.

Flow cell has a liquid sheath fluid in the form of liquid stream that helps in carrying and aligning cells so that they could easily pass as single file via light beam for sensing. Measuring system commonly uses measurement of conductivity or impedance and optical systems including lamps of mercury and xenon; water-cooled high-power lasers *viz.* argon, dye laser, krypton; air-cooled low-power lasers *viz.* red-He Ne (633 nm), argon (488 nm), He Cd (UV) green-He Ne, diode lasers (green, red, blue and violet) resulting in light signals. The detector and analog-to-digital conversion (ADC) system interchanges and converts analog measurements of forward-scattered light (FSC) and side-scattered light (SSC) as well as fluorescence signals specific to dye into digital signals that can easily be processed by a computer. "Acquisition" is the term given to the process of collection of data from the samples through flow cytometer and can be done with the help of the computer connected physically with the flow cytometer and having all the software which can handle the digital interface along with the hardware i.e. computer. Parameters *viz.* voltage and compensation etc. can easily be handled by the software. Initially flow cytometer are used only for experimental research purposes; but with the advent of scientific intervention now it has been routinely used in diagnosis and research leading to the development of fluorescently labeled antibodies.

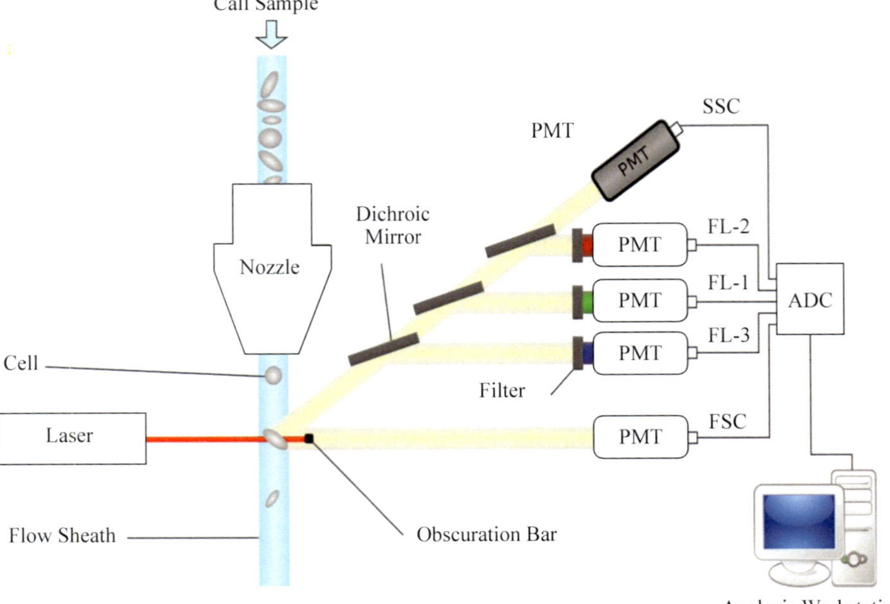


**Figure 1:** Sheath focusing to data acquisition; A schematic diagram of a flow cytometer adapted from work of Kierano.

## Basic Principle of Flow Cytometry Process

- Single cell suspension is entrained in the center of a narrow, rapidly flowing stream of liquid.
- Flow is arranged in order to separate the cells on larger area as compared to diameter.
- Stream of cells are broken down into individual droplets with the help of vibrating mechanism.
- Prior to breaking of stream into droplets, the flow entrains via a fluorescence measuring work station so that fluorescent character of interest of each cell is measured perfectly.
- An electrical charging ring should be placed where stream turns into droplet.
- A positive and negative charge should be placed on the ring and droplet; respectively.
- Based upon their charge, due to electrostatic deflection system , these charged droplets fall into containers.
- Stream is then returned to neutral after the droplet breaks off.
- An fluorescent based antibody specific to a particular cellular protein is added to single cell suspension sample.
- Specific cells pass through a laser beam are monitored for florescence.
- Based on fluorescently-tagged antibody, droplets containing a single cell are assigned a positive or negative charge.

# 13

# Determination of Aflatoxin Using HPLC

*Shalini Vaswani and Mukul Anand*

High-performance liquid chromatography or high-pressure liquid chromatography (HPLC) is a chromatographic method that is used to separate a mixture of compounds in analytical chemistry and biochemistry so as to identify, quantify or purify the individual components of the mixture. It involves the injection of a small volume of liquid sample into a tube packed with tiny particles (1.8- to 5-micron (μm) particle size called the stationary phase) where individual components of the sample are moved down the packed tube (column) with a liquid (mobile phase) forced through the column by high pressure delivered by a pump. These components are separated from one another by the column packing that involves various chemical and/or physical interactions between their molecules and the packing particles. These separated components are detected at the exit of this tube (column) by a flow-through device (detector) that measures their amount. An output from this detector is called a "liquid chromatogram".

Aflatoxins are toxic and carcinogenic compounds produced as metabolites by many Aspergillus species. This fungus invades habitats such as soil, grain, nuts, or decaying vegetation when- ever suitable conditions such as high moisture and temperature are met. B1 is the most toxic of at least 14 different types of naturally occurring aflatoxins. Therefore, many countries and regulatory agencies impose strict limits on B1 as well as other common aflatoxins such as B2, G1, and G2 despite exhibiting lower toxicity. The U.S. FDA set the action level to20 μg/kg (ppb) for human food and to 300 μg/kg (ppb) for animal feed. Aflatoxin $B_1$ is extremely widespread and is present in food and feed products such as cottonseed, corn, and peanuts. It is highly toxic and has been classified as a group 1 carcinogen by the WHO. The aflatoxins $B_2$, $G_1$, and $G_2$ are usually found accompanying $B_1$, in lower concentrations in the contaminated samples. The order of toxicity is AFB1 > AFG1 > AFB2> AFG2.

The determination of aflatoxins is usually done by HPLC with fluorescence detection. However, due to the quenching of the fluorescence activity of some aflatoxins in aqueous mobile phases, techniques are developed to derivatize aflatoxins into nonquenchable forms. The most commonly used technique is UV light derivatization method. Aflatoxins B1, and G1, do not fluoresce under standard reverse phase HPLC conditions unless derivatized by chemical or photochemical means. Post-column derivatization of aflatoxins can increase detectability and/or selectivity of responses for the HPLC detector. By performing the derivatization photochemically, the derivative structures B2 and G2 are apparently obtained, providing the enhanced signals for the B1 and G1 aflatoxins without effect on the B2 and G2 aflatoxins.

## Instrument and Reagents Used

The Agilent 1260 Infinity LC System with Quaternary Pump, Autosampler, thermostatted Column Compartment and Fluorescence Detector equipped with standard flow cell is used for detection.

*Post column UV derivatization module*: UVE UV Derivatization Module for the Analysis of Aflatoxins. This module is connected to the flow path between the column and the fluorescence detector (FLD). PHRED/PROD is a compact unit placed between the HPLC column and the detector to perform on-line post-column continuous photolytic derivatization to increase the sensitivity and/or selectivity of response of fluorescence, ultraviolet, electrochemical, and chemiluminescence detectors.

*Column used*: Agilent Poroshell 120 EC-C18,4.6 x 50 mm, 2.7 µm, The immunoaffinity column for sample preparation (RIDA Aflatoxin column) (from Biopharm Co.)

*Software*: Agilent Open LAB CDS ChemStation

*Solvents and samples*: Methanol (HPLC Grade), Ultra-pure water is prepared using a MilliQ system from Millipore Co.

*Standards*: The aflatoxin standard (containing B1: 2.0 µg/mL, B2: 0.5 µg/mL, G1: 2.0 µg/mL and G2: 0.5µg/mL in acetonitrile (10ml) is purchased from Trilogy.

## Chromatographic conditions

| **Column** | **Agilent Poroshell 120 EC-C18, 4.6 x 50 mm, 2.7 µm** |
|---|---|
| Mobile phase | Water/Methanol 65:35 (v:v), isocratic |
| Stop time | 17.0 min |
| Flow | 1.0 mL/min |
| Injection volume | 30 µL |
| Column temperature | 30 °C |
| FLD | Ex.: 365 nm, Em.: 440 nm, Data rate: 1.16 Hz, Gain: 10<br>Standard flow(8 µL) |

## Procedure

For preparation of standards, 50-μL standard solution is transferred into a 1-mL vial The compounds are redissolved in 950 μL methanol/water (50:50, v:v) and diluted to the concentrations 0.25, 0.5, 1.0, 2.0 and 5.0 ppb.

For sample preparation, 5g of the sample and 0.5g sodium chloride is weighed into a 50-mL centrifugal tube and 25 mL of HPLC grade methanol/water (70:30, v:v) is added. After high speed homogenization for 1 minute and centrifugation at 5,000 rpm for 5 minutes, exactly 5 mL of the supernatant is added to 15 ml of water. The entire sample (20 ml) is passed through the immunoaffinity column or cleanup with the immunoaffinity column and the following protocol is used:

1. Rinse the column with 2 ml distilled water.
2. Pass sample extract solution slowly and continuously through the column(1 drop/second), use positive pressure with syringe, discard the eluent.
3. Rinse the column with 10 ml distilled water, discard the eluent.
4. Air purge the column for about 10 seconds to remove remaining eluent.
5. Elute with 0.5 ml methanol (100%) with a low flow (1 drop/second) rate through the column to ensure complete elution of the aflatoxins.
6. Collect all remaining eluent by purging air through the column.
7. Add 0.5 ml distilled water to the eluent and mix with a vortex mixer.
8. Filter sample through a 0.22 μm syringe filter.
9. Now, run the blank, standards and sample in HPLC using the chromatographic conditions as mentioned above. Figure 1 shows the baseline separation of four aflatoxins in 8.5 min utes. The relatively broad peaks are caused by the volume of the reaction coil inside the UV Derivatization module required to ensure long enough expo- sure of the compounds to the UV light.
10. Five concentrations of the standard (level 1–5) are injected. The linearity (correlation factor) for all four aflatoxins is less than 0.99, and record the instrument reading after data analysis.
11. Then, place the instrument reading in the formula below and get the concentration of aflatoxin in the sample.

## Calculation

$$\text{Instrument reading X} \frac{\text{Sample volume(ml)}}{\text{Sample wt. (g)}} \text{X} \frac{\text{Final Volume (ml)}}{\text{Final sample volume}}$$

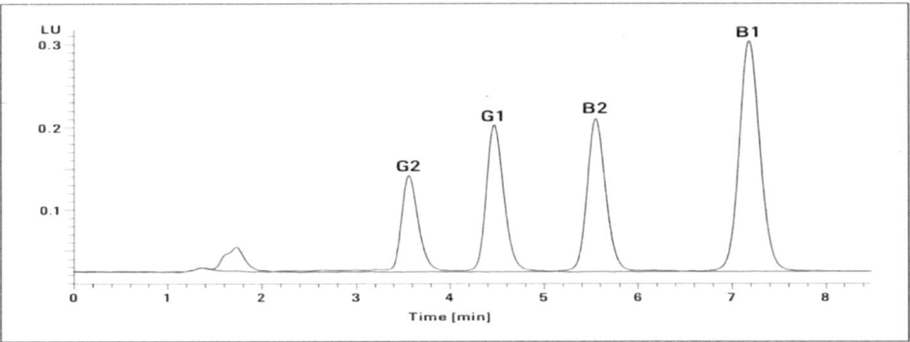


**Figure 1:** Blank injection (red) and analysis of standards (blue).

## Precautions

1. Use clean borosilicate glass bottle only.
2. Rinse bottle with desired solvent before refilling it.
3. Bottles can get contaminated with detergents
4. Algae growth may block the degasser or filters.
5. Precipitation of insoluble salts may block filters or capillaries.
6. Select solvent volume to be used up within 1–2 days.
7. Use only HPLC-grade solvents and water filtered through 0.2 μm filters.
8. Residues or contaminations may block filters or capillaries.
9. Label bottles correctly with bottle content, and filling date/ expiry date.
10. Reduce risk of algae growth: use brown bottles for aqueous solvents, avoid direct sunlight or wrap the bottles in aluminium foil.
11. Flush all channels with water at 2.5–3 mL/min for 5 min to remove salt deposits if buffer applications are used.
12. Use columns only in the marked direction.
13. Always use suitable fittings for your specific column.
14. Flush the column with appropriate solvent found in the column manual.
15. Remove and seal column, and store according to good laboratory practice if needed.
16. Avoid unnecessarily switching on/off the lamp of detector
17. Flush the flow cell after use, Use HPLC grade water to remove salts and use isopropanol to remove organic solvents.
18. Before removing an unused flowcell for storage fill it with isopropanol to prevent algae growth.

## Suggested Readings

Action Levels for Poisonous or Deleterious Substances in Human Food and Animal Feed. Industry Activities Staff Booklet. U.S. Food and Drug Administration, Washington, DC, 2000. http://www.cfsan.fda.gov/~lrd/ fdaact.html

Application of Immunoaffinity Columns for Sample Clean-up Prior to the HPLC Analysis for Aflatoxins. Instruction of AFLAPREP® Immunoaffinity Columns provided by R-Biopharm Rhône Ltd. http://www.r-biopharm.com.

Determination of Aflatoxins (B1, B2, G1, and G2) in Corn and Peanut Butter by HPLC-FLD Using Pre- column Immunoaffinity Cleanup and Post-Column Electrochemical Derivatization", Application Note, Publication number 5990-9125EN, 2011.

# 14

# ICP-OES to Estimate Minerals in Biological Samples

*Shalini Vaswani and Mukul Anand*

## Introduction

Minerals are present in all body tissues and fluids to carry out the physicochemical processes of life; both in animals and human. It is broadly classified into two categories: (1) macro and (ii) micro or trace elements. The macro elements include calcium, phosphorus, potassium, magnesium, sulfur, sodium and chloride, while elements such as iron, copper, cobalt, iodine, zinc, manganese, molybdenum, selenium and chromium come under micro elements. Animals need large amounts of calcium for construction and maintenance of bone and normal function of nerves and muscles. Phosphorus is a vital constituent of adenosine triphosphate (ATP) and nucleic acid and is also essential for acid-base balance, bone and tooth formation. Iron is an important component of the cytochromes for carrying out cellular respiration in addition to its vital role in red blood cells. Magnesium, copper, selenium, zinc, iron, manganese and molybdenum are important co-factors found in the structure of certain enzymes and are indispensable in numerous biochemical pathways. Animals need iodine to make thyroid hormones. Sodium, potassium and chlorine are important in the maintenance of osmotic balance between cells and the interstitial fluid. Excessive intake of some minerals can upset homeostatic balance and cause toxic side effects.

The determination of minerals in feed, environmental and biological samples is challenging due to the wide range of concentrations present, which may vary from ppb (ng/g) to ppm (pg/g) to percent (%) levels. Due to the wide variations of mineral levels,a multi-element technique must be able to simultaneously monitor trace levels of elements in the presence of macro levels of other elements. Inductively coupled plasma-optical emission spectrometry (ICP-OES) is a powerful tool for the determination of metals in a variety of different sample matrices. The ICP-OES features multi-element capability, wide linear dynamic range, high analytical sensitivity, and high sample throughput.

## Principle

In ICP-OES, the sample is heated at temperature level for atomization and during this phase a significant proportion of the atoms are in an excited state. Atomic emissions are produced when the electrons in an excited state fall back to lower energy levels. Since the allowed energy levels for each atom are different, they each have characteristic emission spectrum from which they can be identified. The emitted radiation is passed through a wavelength selector to isolate specific peaks in the spectra corresponding to the atom of interest, and the intensity of the peak is measured using a detector and displayed on a read-out device. An ICP-OES uses high energy argon plasma to convert elements in a solution into a gaseous, excited state form that emits electromagnetic radiation at characteristic wavelengths. The colours of the emitted light and the light intensity can be used to identify the element and determine how much of the element is present in a sample. The ICP-OES uses an array detector so that many elements in a sample can be determined simultaneously.

The ICP-OES is composed of two parts: the inductively coupled plasma (ICP) and the optical emission spectrometer (OES). Argon gas is typically used to create the plasma. An aqueous sample is converted to aerosols via a nebulizer. The aerosols are transported to the inductively coupled plasma (coil of the (RF) generator surrounds part of the quartztorch) which is a high temperature zone (8,000-10,000°C). The analytes are heated (excited) to different (atomic and/or ionic) states and produce characteristic optical emissions (lights). These emissions are separated based on their respective wavelengths and their intensities are measured (spectrometry). The intensities are proportional to the concentrations of analytes in the aqueous sample. The quantification is an external multipoint linear standardization by comparing the emission intensity of an unknown sample with that of a standard sample.

## Procedure

The analysis of minerals in ICP-OES involves following two major steps:

i. Sample digestion
ii. Analysis by ICP-OES instrument

## Digestion of Samples for Preparation of Mineral Extract

The main purpose of sample digestion is to separate out the minerals from the organic matter. The commonly used methods of digestion are dry ashing, wet digestion and microwave digestion.

## Dry Ashing

- A suitable quantity of feed sample (approximately 2.0g) is accurately weighted in a cleaned and dried empty porcelain crucible.
- Then it is charred by keeping it on a heater, until the fumes disappear.
- Charred sample is then transferred to a temperature-controlled Muffle furnace for 4-5 h by maintaining the temperature at 500-600°C for complete ashing till no carbon remains.
- After complete ashing, Muffle is allowed to cool and sample is taken out.
- Approximately 10-15 ml of 5N HCl is added and kept on a hot plate until the volume gets reduced to 1/3rd volume.
- Little water is added to dissolve the salts and then filtered in to a volumetric flask using Whatman filter paper No. 41 and the volume is then made up to desired level with Millipore water.
- The mineral extract thus prepared is ready for mineral estimation.

## Wet Digestion

- Accurately a suitable quantity of sample is taken into 500ml digestion flask.
- 20-30 ml of con HNO3, is added and is boiled gently till white evolved.
- Solution is cooled and 10 ml of 70% HCLO is added.
- Gently boiled by adjusting flame, until solution becomes colorless.
- Solution is then cooled and 50 ml of water is added.
- Then it is boiled to drive out any remaining NO, fumes, cooled, diluted and filtered into a volumetric flask using Whatman filter paper No. 41.

(Note: Care should be taken such that boiling should not be carried till the tubes get dried, it may cause explosion)

## Microwave Digestion

There are several advantages of microwave digestion. It avoids loss of volatile elements such as As, Pb, Se, Hg etc. It requires less sample quantity and time saving process. In this method, the acid digestion of sample is performed in closed vessel by using temperature and pressure-controlled microwave heating system. The temperature and pressure need to change according to sample matrix which could be saved in the program.

- Weigh accurately a suitable quantity of sample (approximately 0.5g) in PTF vessel

- Add 6-8ml of $HNO_3$, (65%) and 1-2ml of $H_2O_2$, (30%) drop by drop and then gently swirl the solution to homogenize the sample with the acids
- Close the vessel and put into the rotor assembly
- Insert the rotor assembly into the microwave chamber
- Run the microwave program already loaded in the method menu or create a new method
- After completion of digestion and cooling, take out the rotor assembly from microwave chamber
- Open the vessel and add little water and then filter in a volumetric flask using Whatman filter paper No. 41 and make up the volume with Millipore water,

## Preparation of Standard Solutions

### Stock Standard Solutions

Commercially prepared single element solutions prepared specifically for spectrometric analysis (usually 1000 or 10,000 mg/L). Stock standard solutions may also be prepared in the laboratory from high purity (>99.99%) metals or salts. Alternatively, commercial multi-element solutions prepared specifically for spectrometric analysis can be used.

### Working Standard Solutions

Three calibration standards are prepared from certified stocks using serial dilution to provide three orders of magnitude calibration in the expected concentration range for each element. Both single element and multi-element stocks are used to create the final element cocktail and the acid strength is matched to the diluted digest solutions above. A recommended maximum concentration of an element in a standard solution is 10 mg/L. Exceptions would be elements usually present at high concentrations for example, calcium, sodium, potassium, magnesium and phosphorus. All aqueous solutions and dilutions are prepared with ultrapure MilliQ water. The range of standards used depends on the elements to be analyzed. The following standards are recommended for various elements.

### QC Standard - Instrument Calibration Verification Standard

Prepare a combined metals standard having concentrations near the midpoint of the calibration curve, but different from those used in any calibration standard. Prepare in same manner calibration standards are prepared, diluting with 2% HNO3.

### Standard Blank

2.0% nitric acid. Prepare sufficient amount for use in standardization determination of IDLs, and for nebulizer rinse between each measurement.

### Instrumental Settings - ICP-OES Analysis

- Adjust the instrumental parameters of the ICP-OES system in accordance with the manufacturer's manual.
- About 30 min prior to measurement, adjust the instrument to working condition.
- Program instrument method for the analytes of interest.
- Suggested emission line wavelengths
- Use background correction.
- Program instrument to use a linear, least squares calculated intercept, curve fit algorithm for converting emission values to mg/L concentration units.
- Follow manufacturer's recommendations for optimizing the emission spectrometer
- After instrument warm-up, perform optical profiling with a 1 mg/L Mn solution.
- Check the sensitivity and the stability of the system.

The samples are analyzed by an Inductively Coupled Plasma Optical Emission Spectrophotometer (ICP-OES) using either axial or radial viewing of plasma depending on the elements. A method is set up to give optimum resolution settings for each element taking into consideration spectral interferences and expected analyte levels. The operational conditions, the analytical lines used, measurement parameters and the wavelengths of the elements are shown below.

The sensitivity and the optimum and linear ranges for each element will vary with the wavelength, spectrometer, matrix, and operating conditions. Some analysts, however, may require that the analyst deviate from the standard operating conditions.

### Table 3 Instrument Parameters

### Parameter

- Nebulization gas flow rate 0.55 L/min
- Auxillary gas flow rate 0.2 L/min
- Plasma gas flow rate 15 L/min

- Sample flow rate 1.5 ml/min
- Operating power (RF) 1300 W
- View Axial/Radial
- Sample volume uptake rate1.0 ml/min
- Spray chamber Cyclonic
- Nebulizer type Cross flow
- Replicates 2

## Calibration of ICP-OES

Before a sample is analyzed and the elemental concentrations of interest can be determined, the instrument must be calibrated. This is done by creating a calibration curve for the measurement of emission for standard solutions and a blank. Calibration of the instrument is done by obtaining standard curve for the respective elements; Multi-element calibration standard solutions are prepared from single and multi-element primary standard solutions.

## Quality Control

1. Following calibration, analyze one high instrument calibration standard, one instrument calibration verification standard and one quality control sample.
   a) Instrument Calibration Standard: Variation of values must be within 2-3% of the known value.
   b) Instrument Calibration Verification Standard: Variation of values mustbe within 10% of the certified values.
   c) Quality Control Sample: Values for all elements must be within limits establwashed
2. Analyze a high instrument calibration standard after each tenth sample and at the end of the set of samples.
   a) Variation of values must be within 8% of the known values.
   b) If any of the values are greater than 8% from the known values, recalibrate the instrument and begin sample analysis from the last"good" instrument calibration standard.
3. Prepare one duplicate sample for each 10 samples. If the set contains less than10 samples, prepare one duplicate per set.
   a) Results on the duplicate sample should agree within 20% of theaverage value of the two samples.
4. Recovery- To avoid contamination and/or loss of elements during sample preparations and measurement methodology, a recovery test is

demonstrated by standard addition methodology. It is performed using a bulk sample which has also been through all digestion procedures. A multi-element standard solution spike is added to a known amount of the samples of normal subjects both before and after digestion, to assess the validity of the digestion procedure. The recoveries of the pre digestion (pre-spiked) sample and the recoveries of the post-digestion (post-spiked) sample are to be determined. The recovery of 90 indicates that there is no contamination and/or loss of elements during sample preparations and measurement steps. The accuracy observed to be quite satisfactory. The percentage recoveries of selected elements in

samples of normal subjects are calculated according to equation:

% Recovery - [Xs/ (Xu. K)] x 100

Where: Xs - measured mean value for spiked sample; Xu - measured mean value for unspiked sample; K - known value for the spike in the sample.

## Data Analysis and Calculation

The quantitative values must be reported in appropriate units, such as micrograms per liter (ug/L) for aqueous samples and milligrams per kilogram (mg/kg) for solid samples. If dilutions are performed, the appropriate corrections must be applied to the sample values. All results should be reported with up to three significant figures.

If appropriate, or required, calculate results for solids on a dry weight basis as follows:

1. A separate determination of percent solids must be performed.
2. The concentrations determined in the digest are to be reported on the basis ofthe dry weight of the sample.

   Concentration (mg/kg) = C * V* D/W

Where,C = Analyte concentration in final extract (mg/L)

V - Final volume in litre after sample preparation

D - Dilution factor (Diluted volume/aliquot volume), if secondary dilution is made

W-Weight in kg of dried sample

## Operation Procedure (ICP-OES Model: Agilent 5800)

1. Check the argon gas pressure in the cylinder
2. Adjust the argon and gas flow pressure as per the instrument's requirement
3. Switch on the air compressor to build air pressure

4. Switch on the chillers and wait for reaching the desired temperature (-20°C)
5. Lock peristaltic pump platens down
6. Switch on the instrument
7. Check the pump for working properly and confirm that the iste is being pumped out
8. Confirm that the exhaust fan is on
9. Open the software in computer
10. Click to open the method for analytes of interest
11. Select the wavelengths for specific analyte
12. Set the method for instrument parameters
13. Select the calibration blank and standards in calibration window
14. Choose the standards unit and sample unit
15. Open sample information window and give details of the sample like sample weight, volume made, aliquot and dilution factor and then save the sample information
16. Click to open the worksheet followed by new worksheet.
17. Ignite the plasma and warm up for at least 30 minutes
18. Aspire 2.0% nitric acid for one to two minutes to clean the system
19. Run the blank and standards for standardization
20. Check and confirm the standardization result by checking the slopes, intercepts, and correlation coefficients
21. Run the samples
22. After every 10 samples check the sensitivity with standard test sample. If any deviation, calibrate the standard curve again

**Shutdown Procedure**

- Thoroughly rinse the sample introduction system by aspiring 2% nitric acid for 5minutes and then by aspiring deionized water for 5-10 minutes
- Extinguwash the plasma
- Close all opened windows and worksheets
- Close the software in computer
- Switch off the instrument
- Release the peristaltic pump
- Switch off the air compressor and chillers

- Turned down the argon gas flow
- Close the gas cylinders
- Switch off the exhaust fan
- Switch off the main power supply

## Instruments and Apparatus

1. Inductively-coupled argon plasma spectrometer
2. Analytical balance
3. Whatman No. 41 ashless filter paper (or equivalent)
4. Volumetric flasks, beakers, and pipettes as required for preparation of reagents and standard solutions

## Reagents

For the determination of elements at trace and ultra-trace level, the reagentsshall be of adequate purity. The concentration of the analyte or interfering substances in the reagents and the water should be negligible compared to the lowest concentration to be determined.

- Millipore Water
- Multi element Standards
- 5N Hydrochloric acid
- Concentrated Nitric acid

## Precautions

- Use hand gloves and face mask while handling acids.
- Sampling must be done carefully and a small and representative portion from the large sample in such a way that any subsequent test on the sample will give a reproducible value.
- Precautions are to be taken to avoid losses by volatilization of elements during ashing.
- Concentrations of mineral elements in feeds are often at the trace level and so it is important to use very pure reagents when preparing samples for analysis.
- Similarly, one should ensure that glassware is very clean and dry, so that it contains no contaminating elements.
- All glassware and plastic ware are washed with deionized water, soaked in 2% HNO3, overnight, rinsed with deionized water, and air-dried.
- It is also important to ensure there are no interfering substances in the sample whose presence would lead to erroneous results.

- Many metal salts are extremely toxic if inhaled or swallowed. Extreme care must be taken to ensure that samples and standards are handled properly and that all exhaust gases are properly vented. Wash hands thoroughly after handling.
- Concentrated nitric and hydrochloric acids are moderately toxic and extremely irritating to skin and mucus membranes. Use these reagents in a hood and if eye or skin contact occurs, flush with large volumes of water. Always wear safety glasses or a shield for eye protection when working with these reagents. Hydrofluoric acid is a very toxic acid and penetrates the skin and tissues deeply if not treated immediately.

# 15

# Recording of Myometrial Reactivity Using Polyphysiograph

***Soumen Choudhury, Amit Shukla, Preeti Singh and Atul Prakash***

Isolated tissue preparations are widely used in physiological or pharmacological research to delineate the mechanism of action of drugs or any active biomolecules. Traditionally it is used as a tool for bioassay of drugs. The tissue used for such pharmacodynamics studies may be isolated from different animal models *viz.* frog, rodents, rabbits, sheep, goat, buffalo, chicken etc. Evaluation of drug-receptor interaction and underlying mechanism of action is then performed using isolated organ bath or myograph where the isolated tissues are mounted under optimum resting tension and changes in the tension following exposure of drug is recorded with the help of recording device. With the advancement of science and technology, the recording devices are also gradually upgraded from smoked drum to student's physiograph to the most advanced Data Acquisition System (DAS) based polyphysiograph.

The isolated tissue preparations (uterine strips, arterial rings, tracheal chains etc.) need essential and appropriate nutrients and environment during experimental which is generally provided by using suitable salt solutions (Physiological Salt Solutions) which provide ionic balance, energy supply, maintain pH and tissue tonicity. The composition of physiological salt solutions is such that it provides an artificial media simulating with the inorganic composition of blood plasma together with a buffer mechanism to maintain an optimum pH of 7.0 to 7.4 and glucose to facilitate tissue metabolism. Commonly used PSS are Frog ringers, tyrode, Kreb's, de Jalon, Ringer-Locke, modified Kreb's Henseleit solution etc.

## Isolated Uterine Preparation

Myometrium is a phasic smooth muscle that exhibits spontaneous and agonist-induced contractions. The rhythmicity and generation of such contractions is intimately related to the generation of slow waves and superimposed action potentials (Inoue *et al.*, 1990). The excitability of tissue, in part, is governed by the resting membrane potential which is determined by opposing inward

($Na^+$, $Ca^{2+}$) and outward ($K^+$, $Cl^-$) ionic fluxes. Individual myocytes may display pacemaker activity but unlike cardiac myocytes, the activity is not confined to particular specialized cells; rather is mobile and this has, therefore, hindered studies into the ionic conductance underlining such activity. From our lab we have reported the existence of excitatory alpha and inhibitory beta adrenoceptors in buffalo myometrium. Further cellular coupling of potassium channels ($K_{ATP}$ and $BK_{Ca}$) with beta$_2$-adrenoceptor has also been documented (Choudhury *et al.,* 2010). Recently we have reported the calcium influx and release mechanisms of histamine-induced myometrial contraction (Sharma *et al.,* 2014).

## Calibration of Physiograph

The change(s) in tissue tension can be measured with help of a highly sensitive isometric force transducer and recorded in a PC using Lab Chart Pro V6 software programme (AD Instruments, Australia). Sensitivity of the instrument is set as required and the sampling rate is adjusted to 5 samples/sec. After calibrating the bridge amplifier at two points, keep the signal range at 10mV.

## Isolation, Mounting of Myometrial Strips and Recording of Isometric Tension

- Collect the uterus in ice cold PSS. For uterine tissue, Ringer-Locke solution (RLS) of the following composition (mM/L): NaCl, 154;KCl, 5.6; CaC12, 2H2O, 2.2; NaHCO3, 6.0; D-Glucose, 5.55 andhaving pH of 7.4 is preferred.
- Dissect the longitudinal strips (approximately 3-10 mm) from each uterine horn and clean off the surrounding tissues and fat. Immediately place the strips in petri dwash containing chilled cold RLS.
- Assemble the organ bath assembly connected with physiograph as per the experimental conditions.
- Tie both the ends of uterine strip with thread and mount in a thermostatically controlled (37 ± 0.5°C) organ bath (UgoBasile) of 10 ml capacity containing continuously aerated RLS. The optimal passive tension in rat uterine strips is 1.0 g (the optimum resting tension is may vary depending on the tissue can be determined by length-tension relationship).
- Equilibrate the tissue under this resting tension for 30-60min. During this period, change the bath fluid once in every 10 min.
- To assess the response of drug on normal myometrial spontaneity, expose the myometrial strips to different concentrations of spasmogen

or relaxant agents. For example, acetyl choline (10nM - 100μM) can be added cumulatively at 0.5 log dose units to observe the uterotonic effect while Salbutamol/Terbutaline is added cumulatively (0.3 nM -0.03 μM) at 0.5 log dose unit.

- After recording the agonist response, give sufficient washing and allow the tissue to regain its normal spontaneity. To assess the antagonism, incubate the tissue with suitable antagonist (atropine, nifedipine etc.) for 30 minutes and in the presence of antagonist record the response of agonist.
- Elicit the cumulative dose response of mercury/ lead at 0.5 log dose units. After recording the response, give sufficient washing to the tissue so that it regains its normal spontaneity.

## References

Asokan KT, Sarkar SN, Mwashra SK and Raviprakash V. (2002). Effects of mibefradil on uterine contractility. Eur. J. Pharmacol. 455 : 65-71.

Boulanger CM, Morrison KJ and Vanhoutte PM (1994). Mediation by M3-muscarinic receptors of both endothelium-dependent contraction and relaxation to acetylcholine in the aorta of the spontaneously hypertensive rat. Br. J. Pharmacol.112: 519-524.

Cohn HI, Harris DM, Pesant S, Pfeiffer M, Zhou Rui-Hai, Koch WJ, Dorn GW, Eckhart AD. (2008). Inhibition of vascular smooth muscle G protein-coupled receptor kinase 2 enhances $\alpha_{1D}$-adrenergic receptor constriction. Am J Physiol Heart Circ Physiol. 295: H1695-H1704.

Ghosh MN. (2005). In: Fundamentals of Experimental Pharmacology. 3rd Edn.

Hosoda C, Tanoue A, Shibano M, Tanaka Y, Hiroyama M, Koshimizu TA, Cotecchia S, Kitamura T, Tsujimoto G, Koike K. (2005). Correlation between vasoconstrictor roles and mRNA expression of α1-adrenoceptor subtypes in blood vessels of genetically engineered mice. Br. J. Pharmacol. 146:456-466.

McLean AC, Valenzuela N, Fai S and Bennett SAL. (2012). Performing vaginal lavage, crystal violet staining, and vaginal cytological evaluation for mouse estrous cycle staging identification. J. Vis. Exp. 67: e4389.

Saroj V., Nakade UP., Sharma A., Choudhury S., Hajare SW, Garg SK. (2018). Dose-dependent differential effects of in vivo exposure of cadmium on myometrial activity in rats: involvement of VDCC and Ca2+-mimicking pathways. Biological Trace Elements Research. 181(2):272-280.

Waghe PK, Sarath TS, Gupta P, Kandasamy K, Choudhury S, Kutty H S, Mwashra SK, SN Sarkar (2015). Arsenic causes aortic dysfunction and systemic hypertension in rats: Augmentation of angiotensin II signalling. Chemico-Biological Interactions.237: 104–114.

# 16

# Estimation of Oxidative Stress in Hepatic Tissue

*Amit Shukla and Sakshi Tiwari*

The cytological insult due to exposure of different environmental pollutant *viz.* pesticides, herbicides etc. impose great loss to the vitality and health of the individuals. The basic pathophysiological distrubances following the exposure of these deterrents to the different visceral organs leads to dysfunction of these organs. Imbalance between body's antioxidant defense and peroxidation of the lipid layer of cells' causing free radical generation is defined as Oxidative stress. Free radicals are the molecules or entity that contains one or more than one unpaired electron. Pathophysiology of several diseases are associated with the generation of oxidative insult and many xenobiotics associated toxicity are also a result of oxidative damage and generation of reactive oxygen species and reactive nitrogen species. Alzheimer's Disease is one of the fine example of oxidative damage associated neuronal degeneration.

Oxidative stress can be estimated using following parameters:

1. Lipid peroxidation (Paula *et al.*, 2005)
2. Reduced glutathione (Sedlak and Lindsay, 1968)
3. Superoxide dismutase (Madesh and Balasubramanian, 1998)
4. Catalase (Aebi, 1983)

## Estimation of Oxidative Stress in the Hepatic Tissues

### Preparation of Liver Homogenates

Samples of liver should be weighed and 10% tissue homogenate is prepared in ice-cold 100mM potassium phosphate buffer (pH 7.4). The homogenate is centrifuged for 10 min at 3000 rpm at 4°C. The supernatant is collected and store at -20°C and used for different biochemical estimation.

### 1. Lipid Peroxidation

The extent of lipid peroxidation is evaluated in terms of MDA (malondialdehyde) production, determined by thiobarbituric acid (TBA) method (Paula *et al.*, 2005).

### Reagents

1. Normal saline solution (NSS) is prepared by dissolving 8.5 g of sodium chloride in one liter of distilled water.
2. 10 percent trichloroacetic acid (TCA) solution: 10 g TCA is dissolved in distilled water and the volume is made up to 100 ml with distilled water.
3. 0.67 percent thiobarbituric acid (TBA) is Prepared by taking 0.67 g of TBA in 100 ml of distilled water and warmed up for dissolving TBA.

### Procedure

One ml of tissue homogenate is incubated at 37± 0.5° C for 2 hours. To each sample, 1 ml of 10% w/v TCA is added. The mixture is mixed thoroughly and centrifuged at 2000 rpm for 10 min. To 1 ml of supernatant, 1 ml of 0.67 percent w/v of TBA is added and kept in boiling water bath for 10 min, cooled and diluted with 1 ml of distilled water. Blank is made by adding all the reagents except the homogenate suspension. The absorbance is read at 535 nm using double beam UV-VIS spectrophotometer.

### Calculation

Calculation of lipid peroxidation is done using the molar extinction co-efficient (EC) of MDA-TBA complex at 535 nm, i.e., 1.56 x $10^8$ $M^{-1}cm^{-1}$. The amount of lipid peroxidation is expressed as nanomoles of MDA formed per ml packed cells.

$$\text{LPO (nM MDA ml}^{-1}) = \frac{\text{OD}}{\text{EC}} \text{ x } \frac{\text{Total vol. of reaction mixture}}{\text{Amount of Sample taken}} \text{ x } 10^9 \text{ x DF x IT}$$

Where, OD is optical density, EC is extinction coefficient (1.56 x $10^8$ $M^{-1}$ $cm^{1}$), DF is dilution factor (3) and IT is Incubation time (2 hr). Total volume of reaction mixture and amount of sample taken is 3 ml and 1 ml, respectively.

## Reduced Glutathione (GSH)

GSH is determined by estimating free-SH groups, using DTNB (dithiobis-nitrobenzoic-acid) method of Sedlak and Lindsay (1968). For the estimation of GSH, 10% homogenate is made in 0.02 M EDTA.

### Reagents

1. DTNB (0.01 M): 99 mg of DTNB (Sigma) dissolved in 25 ml of methanol.
2. Tris-buffer (1 M, pH 8.9): 121 g of tris base is dissolved in 950 ml of distilled water, pH is adjusted to 8.9 and the final volume is made up to 1000 ml with distilled water.

3. TCA (50%): 50 g of TCA is dissolved in distilled water and volume is made up to 100 ml.

## Procedure

To 1 ml supernatant of homogenate, 0.8 ml of water and 0.2 ml of 50% TCA solution is added and incubated at room temperature for 15 min. This mixture is centrifuged at 3000 rpm for 15 min and 0.4 ml of the supernatant is taken. To it 0.8 ml of 1 M tris buffer is added, followed by 0.2 ml DTNB (0.01 M) and the absorbance is read at 412 nm within 5 min. Reagent blank contained no sample and sample blank is without DNTB.

## Calculation

The concentration of GSH is calculated using the extinction coefficient 13000/M/ cm and the results are expressed as μmol GSH per g tissue.

$$\text{Reduced GSH} = \frac{\text{OD}}{\text{EC}} \times \frac{\text{Total vol. of reaction mixture}}{\text{Amount of Sample taken}} \times \text{DF} \times 100$$

## Catalase

Catalase (CAT) activity is assayed by a spectrophotometric method as described by Aebi (1983). Reagents: 1) Phosphate buffer (50 mM pH 7.0) (a) 50 mM KH2 PO4 - 1.37g/ 200 ml (b) 50 mM Na2 HPO4 - 1.42 g/ 200 ml Solutions (a) and (b) are mixed in a ratio of 1:1.5 (v/v) and the pH is adjusted to 7.2) H2 O2 (10 mM): 0.1 ml of 30% H2 O2 is diluted to 100 ml in water.

The solution is Checked at 230 nm and the concentration is adjusted using the molar extinction coefficient.

## Procedure

In a test tube 1.99 ml phosphate buffer and 10 μl homogenate are added and the contents are transferred to the cuvette. The reaction is started after adding 1 ml of $H_2O_2$ directly into the cuvette. The optical density is recorded at every 30 sec for 3 min at 240 nm against distilled water (blank).

## Calculation

The activity of catalase is expressed as millimole $H_2O_2$ utilized/min/mg protein and calculated using the formula.

$$\text{Catalase} = \frac{\Delta\text{OD}}{0.067} \times \frac{\text{Total vol. of reaction mixture}}{\text{Amount of Sample taken}} \times \frac{1}{\text{mg of protein}}$$

Wherein, ΔOD: change in absorbance per minute, i.e. either from 30 sec to 90 sec or 60 sec to 120 sec or 90 sec to 150 sec or 120 sec to 180 sec; total volume of reaction mixture = 3 ml; amount of sample taken = 1 ml.

## Superoxide Dismutase

Superoxide dismutase (SOD) is estimated as per the method described by Madesh and Balasubramanian (1998). It involves generation of superoxide by pyrogallol auto-oxidation and the inhibition of superoxide dependent reduction of the tetrazolium dye MTT [3-(4-5 dimethyl thiazol 2-xl) 2, 5-diphenyl tetrazolium bromide] to its formazan, measured at 570 nm. The reaction is terminated by the addition of dimethyl sulfoxide (DMSO), which solublizes formazan crystals. The colour evolved is stable for several hours and results are expressed as SOD Units/mg of protein (one unit of SOD is the amount (g) of protein required to inhibit the MTT reduction by 50 %).

## Reagents

1. Pyragallol (100 μM): 6.3 mg of pyragallol (Sigma) is dissolved in 5 ml of distilled water. 1 ml from this solution is diluted to 100 ml with distilled water.
2. MTT (1.25 mM): 2.58 mg MTT (Sigma) is dissolved in 5 ml of distilled water.
3. Phosphate buffer saline (PBS): PBS is prepared by dissolving NaCl (8 g), KCl (0.2 g), KH2 PO4 (0.2 g) and Na2 HPO4 (0.94 g) in 800 ml distilled water. The pH is adjusted to 7.4 and the volume is made up to 1 liter with distilled water.

## Procedure

The reagents are added in the sample, control and the blank as shown below. The absorbance is read at 570 nm against distilled water (blank).

**Table:** Reaction set up for determination of SOD

| | **Sample** | **Control** | **Blank** |
|---|---|---|---|
| PBS | 0.65 ml | 0.65 ml | 0.65 ml |
| MTT | 30 μl | 30 μl | 30 μl |
| Homogenate | 10 μl | - | - |
| Pyrogallol | 75 μl | 75 μl | 75 μl |
| Sample, control and blank are incubated for 5 min at room temperature | | | |
| DMSO | 0.75 μl | 0.75 μl | 0.75 μl |
| Homogenate | - | 10 μl | - |

## Calculation

Superoxide dismutase is expressed as Units/ mg of protein

Y % = (OD test/ OD control) x 100

SOD (Units/mg of protein) = (mg of protein in 1 ml homogenate/ Y %) x 50 x Dilution Factor

# 17

# Drug Metabolizing Enzymes Assay

*Atul Prakash and Amit Shukla*

## Tissue Collection

Rats are most commonly employed as test subjects to conduct the experiments related to drug metabolizing enzyme assays involving various enzymes i.e. microsomal and non microsomal enzymes. Rats are humanely sacrificed at the end of the exposure period of 28 days of any therapeutic agents or drug and the liver of all the animals has to excise and immediately wash with ice-cold normal saline, weigh and perfuse with ice-cold 1.15% KCl solution containing 0.05 mM EDTA. Livers are processed further for preparation of microsomes for determination of microsomal drug metabolizing enzymes.

## Preparation of Microsomes

Livers tissues are homogenized in 3 volumes (1:3) of ice-cold (1-4° C) homogenization buffer containing 1.15% potassium chloride in 50 mM potassium phosphate buffer solution (pH 7.4). The homogenates are centrifuged at 15000 g for 30 min (Hitachi cs120GX II). The mitochondrial supernatant (8 ml) is transferred to ultracentrifuge tubes and spun at 105000 g for 1 hour in ultracentrifuge to sediment the microsomal pellet. Cytosolic fractions are collected from the middle separating the lipid layer on top in separate micro centrifuge tubes. The microsomal pellet is then washed by adding 8 ml of resuspension buffer (homogenization buffer containing 0.05 mM EDTA) and spun again 105000 g for 15 min. The microsomal pellet so obtained is resuspended in 4 ml of resuspension buffer and 1 ml aliquots are stored immediately at -80° C until assayed for drug metabolizing enzymes.

## Phase I Drug Metabolizing Enzymes

The hepatic microsomes are used for the assay of selected phase I drug metabolizing enzymes *viz.* cytochrome $P_{450}$ and $b_5$ contents, aminopyrine *N*-demethylase and aniline *p*-hydroxylase.

## Cytochrome $P_{450}$

The cytochrome $P_{450}$ content of the microsomal pellet of liver is measured following the method of Omura and Sato (1964). Microsomal suspension is diluted with 0.1M potassium phosphate buffer (pH 7.4) at $4^0$ C. 950µl of diluted microsomes are then added to reference cell and sample cell and are reduced with 50 µl of 0.9 M sodium dithionite to make it 1 ml. After 2 min, baseline is scanned between 450-490 nm. The sample cell is then bubbled with carbon monoxide for 30 sec at the rate of approximately 2 bubbles /sec and the cytochrome $P_{450}$ spectum is then recorded between 450-490 nm. The difference in absorption between 450 and 490 nm is used for the calculation of cytochrome $P_{450}$ content using extinction coefficient of 91cm$^{-1}$ mM$^{-1}$. Carbon monoxide gas is generated freshly by the reaction of formic acid and concentrated sulphuric acid and purified by passing through the potassium hydroxide pellets.

### Calculation

$$\frac{(A_{450-490})\text{ observed} - (A_{450-490})\text{ baseline}}{\text{mg protein / ml of microsomal suspension}} \times \frac{1000}{91}$$

$$\times \frac{\text{Volume of microsomal sample (ml)} + \text{Volume of phosphate buffer (ml)}}{\text{Volume of microsomal sample (ml)}}$$

## Cytochrome $b_5$

The cytochrome $b_5$ content of the microsomal pellet of liver is measured following the method of Omura and Sato (1964). Microsomal suspension is diluted with 0.1 M potassium phosphate buffer (pH 7.4) at $4^0$ C. 1000 µl of diluted microsomes are added to reference cell and 980 µl are added to sample cell. The baseline is scanned between 424 and 490 nm. The sample cell is then added with 20 µl of 11.94 mM NADH solution and mixed by inversion using parafilm and the cytochrome $b_5$ spectrum is then recorded between 424 and 490 nm is used for the calculation of cytochrome $b_5$ content using extinction coefficient of 118 cm$^{-1}$ mM$^{-1}$.

### Calculation

$$\frac{(A_{424-490})\text{ observed} - (A_{4424-490})\text{ baseline}}{\text{mg protein / ml of microsomal suspension}} \times \frac{1000}{118}$$

$$\times \frac{\text{Volume of microsomal sample (ml)} + \text{Volume of phosphate buffer (ml)}}{\text{Volume of microsomal sample (ml)}}$$

## Aminopyrine *N* – demethylase

Aminopyrine- *N*- demethylase is estimated spectrophotometrically by the method of La Du *et al.* (1971). The reaction rate is followed by measuring the quantity of formaldehyde formed.

## Reagents

1. 150 mM potassium phosphate buffer (pH 7.4)
   a) Potassium dehydration orthophosphate (MW=136.09): (1026.66 mg/50 ml).
   b) Dipotassium hydrogen phosphate (MW=174.18): (1306.35 mg/50 ml).
2. 75 mM magnesium chloride (MW= 203.30): (152.47 mg/10 ml)
3. 5 mM NADP (MW= 787.38): (7.873 mg/2ml)
4. 120 mM Sodium isocitrate (MW= 258.1 ): (30.97mg/1ml)
5. Isocitrate dehyrogenase (ICDH) in 1:1 glycerol and water (4.4 U/mg protein)
6. 120 mM Aminopyrine (MW= 231.3): (55.512 mg/2ml)
7. 75 mM Semicarbazide hydrochloride (MW =111.53): (16.73 mg/2ml)
8. 10% Zinc sulphate: (1g/10 ml)
9. Saturated barium hydroxide: Barium hydroxide is added to 10 ml of distilled water till it dissolved.
10. Nash reagent: Freshly prepared (400 µl of acetyl acetone is added to 30 g ammonium acetate and made up to 100 ml. with double distilled water.

## Procedure

To 10 ml test tubes, the enzyme, cofactor and the substrate are added as indicated below:

| | | Test Sample | Blank |
|---|---|---|---|
| a) | PPB | 0.5 ml | 0.5ml |
| b) | $MgCl_2$ | 0.1 ml | 0.1 ml |
| c) | NADP | 0.1 ml | 0.1 ml |
| d) | Sodium isocitrate | 0.1 ml | 0.1 ml |
| e) | ICDH | 10 µl | 10 µl |
| f) | Aminopyrine | 0.1 ml | 0.1 ml* |
| g) | Semicarbazide HCl | 0.1 ml | 0.1 ml |

*Distilled water

The assay mixture is incubated for 15 min at $37^0$ C in orbital shaker to ensure reduction of all NADP to NADPH. Then 0.5 ml of liver microsomal suspension (diluted 1: 1 in resuspension buffer) is added and incubated (SONAR, orbital shaking incubator) for 10 min at $37^0$ C with shaking (120 cycles/min). The reaction is stopped by adding 0.5 ml of 10% zinc sulphate solution followed by 0.5 ml of saturated barium hydroxide solution. Contents are transferred to 5 ml tubes and centrifuged at 2000 rpm for 10 min. Formaldehyde formed is estimated using 1 ml of the clear supernatant. To 1 ml of supernatant from sample and blank, 0.5 ml of Nash reagent is added and vortexed. The tubes are then placed in a water bath at 60 °C for 30 min for color development. The absorbance of the samples and the tissue blank is read against water blank at 412 nm. The amount of formaldehyde formed is calculated after subtracting the tissue blank and using the molar extinction coefficient of 8000.

## Calculation

$$\frac{\text{Absorb. (Test)} - \text{Absorb. (Blank)}}{\text{Extinc coeff.} \times \text{path length (1 cm)}} \times 10^6 \times \frac{\text{Total Vol.}}{\text{Enz. Vol.}} \times \frac{1}{\text{Time of incub. (10 min.)}}$$

$$\times \frac{1}{\text{mg of protein in vol. of enzyme taken}} = \text{nmole formaldehyde formed / min / mg MP}$$

### Aniline *p*-hydroxylase

Aniline p-hydroxylase is estimated spectrophotometrically by the method of La Du *et al.* (1971). Hydroxylation of aniline is followed by determining the rate of formation of the product p-aminophenol.

## Reagents

1. 150 mM Potassium buffer, pH 7.4
2. 75 mM magnesium chloride
3. 5 mM NADP
4. 120 mM Sodium isocitrate
5. Isocitrate dehydrogenase in 50% glycerol (4.4 U/mg protein)
6. 20% Tricloro acetic acid
7. 120 mM Aniline hydrochloride
8. 10% Sodium carbonate
9. 2% alkaline phenol in 0.2N NaOH (2 g alkaline phenol/100 ml of 0.2N NaOH)
10. P-Aminophenol standard (10 mg p- aminophenol/100 ml of 6.67% TCA).

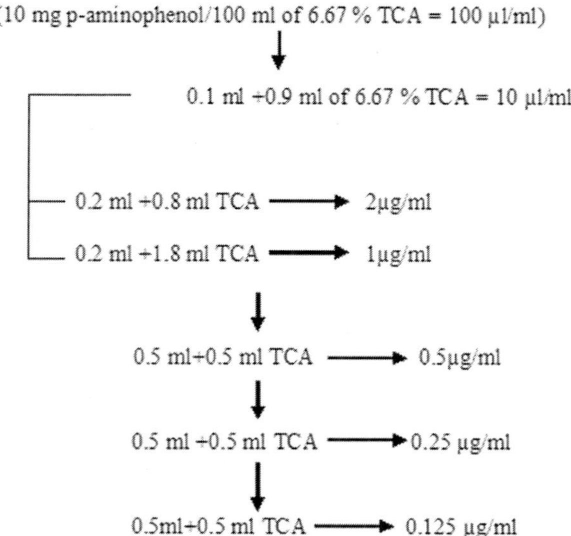


## Procedure

To 10 ml test tubes, the enzymes, cofactor and the substrate are added as indicated below:

a) Phosphate buffer 0.5 ml
b) Mg $Cl_2$ 0.1 ml
c) NADP 0.1 ml
d) Sodium isocitrate 0.1 ml
e) ICDH 10 μl
f) Aniline HCl 0.1 ml
g) Distilled water 0.1 ml

The assay mixture is incubated at 37 °C for 15 min to convert all of the NADP to NADPH. Then, 0.5 ml of microsomal suspension (diluted 1:1 resuspension buffer) is added to start the reaction and incubated (SONAR, orbital shaking incubator) at 37 °C with shaking (120 cycles/min) for 15 min. The reaction is stopped by adding 0.5 ml of 20% TCA. The contents are mixed and transferred to 5 ml tubes and centrifuged at 2000 rpm for 10 min. Reagent blank is prepared using 1 ml of 6.67% TCA. To both TCA supernatant and reagent blank, 0.5 ml of 10% $Na_2CO_3$ is added followed by 1 ml of 2% alkaline phenol in 0.2 N NaOH. The tubes are placed in BOD incubator at 37 °C for 30 min for colour development p- aminophenol reacts with phenol in alkaline medium to form the blue coloured phenol –indophenol complex, which is measured at

630 nm. The amount of p- aminophenol formed is calculated by with standards or using the millimolar extinction coefficient of the indophenol dye in TCA (E630 =28.4 $cm^{-1}$ $nm^{-1}$).

### Calculation

$$\frac{\text{Absorb. (Test)} - \text{Absorb. (Blank)}}{\text{Extinc coeff.} \times \text{path length (1 cm)}} \times 10^6 \times \frac{\text{Total Vol.}}{\text{Enz. Vol.}} \times \frac{1}{\text{Time of incub. (15 min.)}}$$

$$\times \frac{1}{\text{mg of protein in vol. of enzyme taken}} = \text{nmole p} - \text{aminophenol formed / min / mg MP}$$

### Phase II drug metabolizing enzymes

Following enzymes are estimated.

### Glutathione Sulfotransferase

Microsomal glutathione sulfotransferase is estimated by the method of Habig *et al.* (1974).

### Reagent

1. 0.3 M Potassium phosphate buffer (pH 6.5)
   a) Potassium dihydrogen orthophosphate (MW=136.09): (2042 mg/50 ml)
   b) Dipotassium hydrogen phosphate (MW=174.18): (2613 mg/50 ml)
2. 30 mM reduced glutathione (GSH) (MW=307.33): (9.21 mg/ 1 ml)
3. 30 mM CNDB in 95% alcohol (MW=202.55): (60.76 mg/10 ml)

GSH and CDNB are prepared afresh.

### Procedure

To 1 ml of phosphate buffer in a test tube, 0.1 ml of CDNB and 0.1 ml microsomal protein (1:1 dilution in resuspension buffer) are added. 1.7 ml of distilled water is added to make the final volume to 2.9 ml. the reaction mixture is incubated at $37^0$ C for 5 min and the reaction is started by adding 0.1 ml of 30 mM glutathione. The absorbance is measured at 60 sec interval for 5 min at 340 nm. Reaction mixture without microsomal protein is used as blank.

### Calculation

$$\mu\text{mol of CDNB} - \text{GSH conjugate formed / min / mg protein } \frac{\text{OD} \times 3 \times 1000}{9.6 \times 5 \times \text{Protein in mg}}$$

# 18

# Acetylcholine Esterase (AChE) Assay

*Atul Prakash and Amit Shukla*

Acetylcholinesterase activity in brain is measured by the method described by Ellman's *et al.* (1961).

## Reagents

1. Phosphate buffer (0.1M) (pH 8.0)

   **Solution A:** 5.22 g of $K_2HPO_4$ and 4.68 g of $NaH_2PO_4$ are dissolved in 150 ml of distilled water.

   **Solution B:** 6.2 g of NaOH is dissolved in 150 ml of distilled water.

   Solution B is added to solution A to get desire pH 8.0 and then finally the volume is made up to 300 ml with distilled water.
2. DTNB reagent 39.6 mg of DTNB with 15 mg of $NaHCO_3$ is dissolved in 10 ml of 0.1 M phosphate buffer.
3. Acetylthiocholine (ATC) 21.67 mg of Acetylthiocholine is dissolved in 1 ml of distilled water.

## Procedure

The brain tissue is weighed and 10% homogenate is made in 0.1M Phosphate buffer (pH 8.0). Take 0.4 ml aliquot of homogenate in a cuvette containing 2.6 ml phosphate buffer (0.1M) and 100 µl DTNB. The contents of the cuvette are mixed thoroughly by bubbling air and absorbance is measured at 412 nm in a spectrophotometer. When absorbance reached a stable value, it is recorded as the basal reading .To this 20 µl of acetylthiocholine is added and change in absorbance is recorded for a period of 10 min at intervals of 2 min. Change in the absorbance per minute determined.

## Calculation

The enzyme activity is calculated using the following formula and result is expressed in moles of substrate hydrolysed $min^{-1}$ $gm^{-1}$ of wet tissue.

R (moles of substrate hydrolyzed/ min/ gm of tissue) = $5.74 \times 10^{-4} \times A/CO$

Where,

R = Rate in moles of substrate hydrolyzed/ min/ gm tissue

A = Change in Absorbance/ min

CO = Original concentration of the tissue (mg/ml)

# 19

# Preparation of Hepatocyte Culture for Toxicity Testing of Xenobiotics

*S.P. Singh and Atul Prakash*

Pharmaceutical and chemical industries depend largely on toxicological evaluations to provide information regarding safety of their commercial products. Toxicologists so far have been relying on animal studies to investigate potentially beneficial or harmful effects of drugs and chemicals prior to testing them in humans. Traditionally, the safety assessment process during drug development has been based on screening indiscriminately a plethora of new chemical entities (NCEs) using a large number of animals as testing models in order to identify new therapeutic agents with the expectation that one or a few of these NCEs have some competitive advantages over the existing therapeutic agents. That is why they are more efficacious, bioavailable, selective, and safer. This approach is costly, time-consuming, and requires large quantities of test material. Currently, it takes about 10-12 years and a huge expenditure for a drug to move from research laboratory to patent. Thus alternative cell culture models in the early phases of the drug discovery process can save time and money both by eliminating least beneficial NCEs.

## Preparation of Reagents

### Culture media

For the culture of hepatocytes medium M199 (GibcoBRL, USA) is used. One pouch of the dry media is dissolved in 1 liter of filtered distil water. The pH of the media is adjusted by adding 0.4% sodium bicarbonate to be around 7.4. The media is than sterilized by filtration through seitz filter (0.22 mm) and stored in 50 ml sterilized conical flasks.

### Hanks Balanced Salt Solution (HBSS)

For the preparation of HBSS, one vial of dry powder is dissolved and properly mixed in 1000 ml of sterilized distal water. Sterilization is done through filtration. The prepared HBSS may be kept in 100 ml sterilized conical flask for later use.

### Other Accessories

For preparing of tissue culture, tissue culture flask (25 ml), TC plates of 6, 24 and 96 well may be used depending on the parameter to be studied. All the glassware to be used should be properly washed and sterilized.

## Isolation and Preparation of Hepatocyte Cell Culture

### Isolation of Chicken Embryo Liver

For the isolation of chicken hepatocytes, 14-day-old chicken embryos are used. The surface of the eggshell is cleaned with sterilized cotton pads soaked in spirit. Care should be taken to avoid strong movements of the egg to avoid embryonic mortality. The eggshell is then carefully opened using sterilized pair of forceps from the broad end of the egg as air sac of the embryo is situated at this end. After the egg is open, the egg membranes, which cover the embryo, are carefully removed and the embryo is taken out from the egg and put into a sterilized petridish. The embryo is decapitated immediately with a pair of sharp scissors in single stroke. The abdomen cavity is cut open and viscera is shifted to one end and liver carefully removed from the body and put into a petridish having HBSS. After removing the undesirable tissue pieces and gall bladder liver is transferred and washed in petridish with HBSS. After proper washing in HBSS, the liver piece is transferred to a petridish containing sufficient volume of media (M-199). Similar procedure is followed for all the embryos sacrificed. The number of embryos may be used as per the requirement of amount of final cell suspension.

### Isolation of Mammalian Liver Cells

For the isolation of mammalian hepatocytes, new born animals, preferably newly born rat pups are used. The pups are sacrificed by decapitation and the visceral cavity is openend immediately with a pair of sharp scissors, aspecticaly. The viscera is shifted to one end and liver carefully removed from the body and put into a petridish having HBSS. After removing the undesirable tissue pieces and gall bladder liver is transferred and washed in petridish with HBSS. After proper washing in HBSS, the liver piece is transferred to a petridish containing sufficient volume of media (M-199). Similar procedure is followed for all the animals sacrificed. The number of animals used depends on the amount of final cell suspension required.

### Trypsinization

The liver pieces are then passed through the barrel of a 5 ml sterilized fresh syringe to get it minced finely. The minced liver tissue is now subjected to trypsinization. Trypsin solution (1:250) is used for disintegration of the tissue.

About 5 ml of trypsin per tissue piece is sufficient for one step of trypsinization. Trypsinization is carried out two times on the same tissue by changing the trypsin solution. The mixture of tissue in trypsin solution and HBSS is put on magnetic stirrer till tissue debris disintegrates almost completely. The suspension so obtained is then filtered through a muslin cloth tied over a glass funnel.

## Isolation of Hepatocytes

The cellular filtrate obtained after filtration of trypsinized tissue suspension is centrifuged to wash and obtain a pellet of hepatocytes. For this the filtrate is washed twice with HBSS at 3000 rpm for 10 minutes in refrigerated centrifuge, third and final washing is done in media M-199. Pellet, obtained after centrifugation, is mixed in 1 ml of media and the volume of the cell pellet obtained is measured. The final cell suspension is prepared by diluting the pellet 200 times with media. The suspension is observed under the microscope after staining with trypan blue dye to determine the live cell percentage.

## Seeding Rate

The culture plates are then seeded with the required volume of the cell suspension depending on the type of plate used. The concentration of the cells in the cell suspension is kept so as to obtain 1 X $10^4$ cells per ml. The plate or bottle is then incubated at $37^0$C with 5% $CO_2$ pressure in a $CO_2$ incubator for the formation of monolayer of hepatocytes.

## Toxicity Studies

### Test Solution

Technical grade of test compound are used for exposure of the cells in the culture. Test compound is dissolved in suitable solvent to prepare desired dilutions of the compounds. Care is taken to keep the concentration of the solvent less than 0.1% in the culture. Dilutions are made so as to obtain log dose of the compound in the culture. Fresh test solutions are prepared for each round of exposure.

### Exposure to Test Compound

After the formation of monolayer of cells in the culture plate, the cells are exposed to predetermined concentrations of the test compound for a period of 5 hours and the media is washed off. The cells are than incubated with fresh media and observed for cytopathic effects for the next 48 hours. The control plates are exposed to the concentration of solvent present in the test solution.

### Parameters Recorded In Cell Culture

The cultures are examined for mainly morphological changes, viability of cells, cell reproduction, biosynthesis at macromolecular levels, cell transformation-carcinogenesis, mutagenesis and differentiated functions.

### Morphological Examination of the Cells

Cells are examined for any morphological change, either in the culture plate directly under inverted microscope or after preparing smear and staining it with appropriate stain.

### Cell viability Assay

Two methods can be used for viability assay. The cells may be stained with a vital stain and observed for dye exclusion. The estimation of macromolecules like albumin can also be used as an assay for viability of the cells.

### Biosynthesis at Macromolecular Level

The cells in the culture can be estimated for the rate of synthesis of DNA, RNA and proteins as an assay for the toxicity studies. The content of these macromolecules can be estimated in the cells at various levels of growth to determine the time related effect of xenobiotics.

### Enzyme Estimation

The spent media is evaluated for the concentration of lactate dehydrogenase, aspartate aminotransferase and alanine aminotransferase to determine their levels after incubation and exposure to the pesticides. Hepatocytes after exposure to a drug or pesticide or may be used for estimation of parameters relevant in hepatotoxicity studies.

### Mutagenesis

The cells in the culture can be observed for chromosome number and aberrations after treatment with the test compound. Looking for DNA repair and point mutations in the cells can also be used as a parameter for mutagenic effects of xenobiotics.

It is the need of the time that we recognize the diverse need of the society, existing time and fiscal constraints. The scientific community needs primarily the understanding of the mechanisms and also a large number of compounds to be screened. Tissue culture can very well satisfy both the needs of the society and at the same time keep the time and fiscal constraints in view.

## Suggested Reading

Berry, M.N. and Friend, D.S. 1969. High-yield preparation of isolated rat liver parenchymal cells. J.Cell.Biol. 43 : 506-519.

Ekwall, B. 1980. Screening of toxic compounds in tissue culture. Toxicology. 17: 127-142.

Nardone, R.M. 1980. Tissue culture systems in toxicity testing. In A.N.Rowan and C.J.Stratman (Eds), The use of Alternatives in Drug Research, The Macmillin Press, London.

Stammatu, A.P., Silano, V and Zucco, F. 1981. Toxicology investigations with cell culture systems. Toxicology. 20 : 91-153.

# 20

# Detection of Apoptosis By DNA Gel Electrophoresis

*Atul Prakash and Amit Shukla*

## Isolation of Peripheral Blood Lymphocytes

### Chemicals

1. Heparin 100 IU/ml in PBS
2. Sterile PBS (pH 7.2)
3. Histopaque-1077
4. RPMI-1640 Media

### Protocol

Collect 5 ml of blood from the animal in 1 ml (20 IU/ml of blood) of diluted heparin and add 2 ml of PBS. Layer over the diluted blood on 5 ml of histopaque-1077 in 15 ml centrifuge tube and centrifuge at 1500 rpm for 40 min. Collect the buffy coat at the interface of plasma and histopaque in separate centrifuge tubes. Wash the cells thrice with PBS and suspended in 2 ml of RPMI-1640 media supplemented with 10% fetal calf serum. Determine the percentage cell viability by 0.1% trypan blue exclusion test, using hemocytometer. Finally dilute the cell suspension to $10^6$ cells/ml in RPMI-1640 medium.

## Isolation of Splenocytes

### Chemicals

1. PBS (pH 7.2)
2. Histopaque-1077
3. RPMI-1640 Media

### Protocol

Aseptically separate the spleen from the animal in ice cold PBS. Place the spleen on a sterile, autoclaved nylon mash prewetted with PBS. Remove the capsules of the spleen using a pair of forceps and needles. Disintegrate the organ into many pieces with the plunger of a 10 ml glass syringe. Force the tissue through the mesh into a petridish containing chilled PBS. Afterwards pipette the cell suspensions 4-5 times to break large cell clumps. Collect the cell suspensions in 15 ml centrifuge tubes and allow to stand on ice. After 5 min collect the top 12 ml of the suspensions into another set of centrifuge tube and pellet the cell by centrifugation (1500 rpm for 10 min). Resuspend the cell pellets in PBS and layer over histopague-1077 at a ratio of 1:1. Centrifuge the tubes at 1600 rpm for 40 min and collect the interface ring rich in splenocytes. Wash the cells with PBS thrice and adjust the cell concentration to $10^6$ cells/ml in RPMI-growth medium after assessing the cell count and cell viability using trypan blue.

## Cell Viability Count

### Materials and Chemicals

1. 0.1% Trypan Blue Dye (in PBS)
2. Hemocytometer with Cover slips

### Protocol

Take 20 ml of diluted cell suspension in a microcentrifuge tube (MCT) and add 180 ml of 0.1% trypan blue dye. Mix well and immediately charge the hemocytometer, wait for 2 minutes and see under 10X and count the live (bright) and dead (blue) cells in 64 secondary squares (4 large corner square).

### Calculation

1. Average no of cell = Total count/4
2. Total viable cell/ml = (Average no of viable cells) x $10^4$ x 10
3. % Viability = [Average no of viable cell x $10^4$ x 10]/ Total count (Viable + Non-Viable) x 100
4. % Non-Viability = [Average no of Non-Viable cell x $10^4$ x 10)]/ Total count (Viable + Non-Viable) x 100

   **or** (100 - % Viability).

## Extraction of Apoptotic DNA

### Chemicals

1. PBS
2. Lysis buffer

| | |
|---|---|
| i. NP4 (Igepal) (1%) | 100 ml |
| ii. EDTA (20mM) | 400 ml (0.5 M EDTA) |
| iii. Tris-HCl (50 mM) (pH 7.5) | 500 ml (1M Tris) |
| iv. DW | 9.0 ml |

3. RNase A 10 mg/ml
4. Proteinase K 20 mg/ml
5. SDS (10%)
6. Ammonium acetate (10 M)
7. Absolute alcohol
8. Ethanol (70%)
9. Tris-EDTA

## Protocol

Collect the cells in 1.5 ml MCT and pellet the cells by centrifugation at 8000 rpm for 5 min, reconstitute the pellet with 1 ml of PBS, centrifuge again and discard the supernatant. Add 200 ml of lysis buffer to the pellet and vortex. Centrifuge at 4000 rpm for 10 min. Collect the supernatant and again treat the pellet with 100 ml of lysis buffer similarly. Treat the supernatant of both steps with 35 ml of 10% SDS (100 ml of 10% SDS per 1 ml of sample) for 2 hrs at 37°C. Digest with 4 ml of Proteinase K (100 mg/ml of 20 mg/ml conc.) for 3 h at 50°C. Add 175 ml of 10M ammonium acetate (0.5 volume), precipitate with 438 ml (2.5 volume) of absolute ethanol and incubate O/N at –20°C. Centrifuge at 12000 rpm for 15 min. Wash with 1 ml 70% ethanol at 12 K for 5 min. Discard the supernatant and dissolve the pellet in 20 ml of TE and add 1ml RNase A, incubate at RT for 1 h or dissolve the pellet in TE containing RNase A and incubate at room temperature for 3 h. Run the sample in 2% AGE at 60 volts up to half the gel length

## Interpretation

The lane showing ladder pattern is apoptotic, DNA from necrotic cells shows smeared lane while the DNA of the normal cell remains intact near the well of the gel.

## Apoptosis Related Parameters

### Reactive Oxygen Species Generation (ROS generation)

The procedure as described by Bass *et al.* (1983) and modified by Das and Mishra (1994) is followed. Thymocytes are seeded at $10^6$ cells/ml/well in 24-well plates and are exposed to the test compounds at different concentrations for 12 h in 5% $CO_2$ at 37 °C. After the exposure period, thymocytes are collected in microcentrifuge tube; then cells are pelleted at 200 x g for 10 min at 20 °C. Cells are washed with PBS. Dichlorofluorescein diacetate (DCFH-DA) dye at final concentration of 5 μM is added to this sample and again kept for 15 min in the dark at room temperature. Flowcytometric analysis for DCF florescence is done as Counts vs FSC-H and 10000 thymocytes per event are taken. Per cent right shift in the florescence peak is recorded.

### Apoptosis Detection (Flowcytometric Method Using Propidium Iodide Dye)

Thymocytes are isolated aseptically as described elsewhere. After the treatment period, the isolated thymocytes are washed with PBS and fixed drop by drop addition of ice cold 70% ethanol and stored at -20$^0$ C overnight. The fixed cells are washed with PBS twice and suspended in 500 µl of propidium iodide hypotonic buffer and incubated at 37$^0$C for 15 min in dark. The PI fluorescence is measured through FL-2 filter in Becton Dickenson Flowcytometer.

### Mitochondrial Transmembrane Potential (Flowcytometric method using $DiOC_6$ dye)

### Preparation of 3, 3$^1$ Dihexyloxacarbocyanine Iodide ($DiOC_6$) dye solution

40 µM stock solution of $DiOC_6$ is prepared in dimethyl sulfoxide by dissolving 0.292 mg of dye in 1 ml of dimethyl sulfoxide. Ten microliter of the stock solution is diluted to 1 ml using phosphate buffer saline to prepare working solution.

### Protocol

Mitochondrial transmembrane potential is analyzed according to the method of Castedo *et al.* (2002). Briefly, after the treatment period the cells are harvested and washed with PBS and dissolved in 0.5 ml of PBS and 25 µl of $DiOC_6$ is added. Tubes are incubated at 37° C for 10-15 min in dark. After the incubation, the cells are returned to ice and anaylsed in Becton Dickinson flow cytometer through FL 1(emission: 530 nm).

# 21

# Lethal Dose 50 Calculation Methods

*Amit Shukla, Atul Prakash, Soumen Choudhury and Sakshi Tiwari*

**LD 50-** It is defined as the concentration/dose that kills 50 % of the experimental population.

Acute toxic studies have been conducted in order to calculate LD50. There are several methods to calculate LD50, some are as follows:

## Methods to Determine LD 50

- Karber's method.
- Miller & Tainter method
- Lorke method
- Fixed dose method
- Acute Toxic Class method
- Up & down procedure

## Karber's Method

This method involves the administration of different doses of test substance to various groups, which has five animals each. The first group of animals is administered with the vehicle in which the test substance was dissolved or diluted

Second group onward receives different doses of the test substance in increasing order. Observations are recorded (mortality per group) and a table is made.

After preparing table value of LD50 can be calculated by formula -

$LD50 = LD100 - \sum (a \times b) / n$

Where, LD50 = Median lethal dose

LD100 = Least lethal dose

a = Dose difference

b = Mean mortality

n = Group population.

## Miller & Tainter Method

This method involves the administration of different doses of test substance to various groups, which has five animals each. The first group of animals is administered with the vehicle in which the test substance was dissolved or diluted. Second group onward receives different doses of the test substance in increasing order. Observations are recorded (mortality per group) and a table is made .The observed mortality is converted in to percentage mortality and after that the observed percentage mortality is converted into probit. The probit values thus obtained were plotted against log dose on graph.

- The LD50 value and its standard error are determined from the graph
- The dose corresponding to probit 5 is taken as LD50.
- Disadvantage: Too many animals are utilized.

## Lorke's Method

### Phase 1

This phase requires nine animals. The nine animals are divided into three groups of three animals each. Each group of animals are administered different doses (10, 100 and 1000 mg/kg) of test substance. The animals are placed under observation for 24 hours to monitor their behaviour as well as if mortality.

### Phase 2

This phase involves the use of three animals, which are divided into three groups of one animal each. The animals are administered higher doses (1600, 2900 and 5000 mg/kg) of test substance and then observed for 24 hours for behaviour as well as mortality.

Then the LD50 is calculated by the formula:

$LD50 = \sqrt{(D0 \times D100)}$

D0 = Highest dose that gave no mortality,

D100 = Lowest dose that produced mortality.

# 22

# OECD Guidelines for Acute Animal Toxicity Tests

*Amit Shukla and Sakshi Tiwari*

## Introduction

The immense use of chemicals and pharmacological entities has put the public health at brisk of potential health hazards. These substances in modern days are presented in the form of common constituents of food, medicines, beverages and junk foods etc. However, long term consumption of these substances might be responsible for chronic toxicity, but the accidental ingestion of large quantities of these substances may result into adverse acute toxic reactions. Hence an experimenter should have the acumen to develop and perform the acute toxic tests in order to find out the lethal dose 50 i.e. $LD_{50}$ of the substances.

## Acute Toxicity Studies

Acute toxicity studies helps to establish the dose dependent adverse effect including mortalities. The $LD_{50}$ has been used as a major parameter in screening of pharmacological agents for acute toxic potential. Apart from mortality, other biological effects and the time of onset, duration and degree of recovery on survived animals, are also important in evaluation of acute toxicity. Hence, acute toxicity study solely gives information about LD50, therapeutic index and the degree of safety of a pharmacological agent.

Acute toxicity is defined as adverse effects occur following oral or dermal administration of a single dose of a substance, or multiple doses given within 24 hours, or an inhalation exposure of 4 hours. Oral administration is the most common form of acute systemic toxicity testing.

## Objective of Acute Toxicity Studies

- To determine the Median Lethal Dose (LD50)
- To determine Maximum Tolerated Dose (MTD)
- To identify potential target organs for toxicity
- Identify parameters for clinical monitoring
- To help selecting doses for repeated-dose toxicity studies

## Keynotes

- Toxicity is the degree to which a substance (poison) can harm humans or animals
- Toxicology is a science that defines the limits of safety of chemical agents for human & animal populations. (Casarett, 1996)
- All substances are poisons, there are none which is not a poison, the right dose differentiates a poison and a remedy. (Paracelsus saying)

## Why Toxicity Studies?

- Benefit –risk ratio can be calculated
- Prediction of therapeutic index
- Therapeutic index = LD50/ED50 (Smaller ratio, better safety of the drug)
- Toxicity tests are mostly used to examine specific adverse events or specific end points such, skin/eye irritation, carcinogenicity etc.

The US-FDA states that it is essential to screen new molecules for pharmacological activity and toxicity potential in animals first.

## Why Do We Require Non Clinical Studies in Animals Before Administered to Man?

Pharmacological effects are same in man as in animals. Toxic effect in animal species will predict adverse effects in man. Giving high doses in animals improves predictability to man.

## Ethics

Before conducting any toxicological testing in animals or collecting tissue or cell lines from animals, the study should be approved by Institute Animal Ethics Committee (IAEC)

**OR**

The protocol should satisfy the guidelines of the local governing body

## Regulatory Mechanisms in India

- Institute Animal Ethics Committee (IAEC)
- Committee for the Purpose of Control and Supervision of Experiments in Animals (CCSEA)
- Drugs & Cosmetics Act, (1940)
- The Prevention of Cruelty to Animals Act, 1960 as amended up to 30th July 1982 and Animal Welfare Board

## Types of Toxicity Studies

- Single dose studies (Acute toxicity studies)
- Repeated dose studies (sub-acute or Chronic studies)
- Special toxicity studies (Hepatotoxicity, nephrotoxicity etc.)

## Acute Toxicity

- Adverse effects occur following oral or dermal administration of a single dose of a substance, or multiple doses given within 24 hours, or an inhalation exposure of 4 hours. Oral administration is the most common form of acute systemic toxicity testing.

## Objectives

- To determine the Median Lethal Dose (LD50)
- To determine Maximum Tolerated Dose (MTD) and No Observable Effect Level (NOEL)
- To identify potential target organs for toxicity
- Identify parameters for clinical monitoring
- To help selecting doses for repeated-dose toxicity studies

## Acute Toxicity Testing

1. Acute toxicity testing needs two different animal species (one rodent and one non-rodent)
2. Investigational products are administered at different dose levels, and the effect is observed for 14 days
3. All mortalities caused by the investigational product during the experimental period are recorded
4. Morphological, biochemical, pathological, and histological changes in the dead animals are investigated

## Procedure

### 1. Selection of Animal

Preferred sex is female .All of them should be of same sex .After completion of the study in one sex, at least one group of five animals of the other sex will also be tested .Females should be nulliparous and non-pregnant .Each animal, at the start of administration, should be between 8 and 12 weeks old .

### 2. Preparation of Doses

The substance used in the toxicity tests should be as pure as the one intended to be given to humans. Test substances should be administered in a constant volume over the range of doses to be tested, by varying the concentration of the dosing preparation. In rodents, the volume should not normally exceed 1mL/100g of body weight, however in the case of aqueous solutions 2 mL/100g body weight can be considered .

### 3. Administration of Doses

Oral and the peritoneal administration are most commonly used. Administration is made in a single dose by using a stomach tube or an intubation cannula, and when this is not possible, the dose can be given in smaller fraction within 24 hour. Animal should be fastened prior to dosing, in rat food withheld over-night, in mice 3-4 hours with unrestricted water. Following the fasting period the animals should be weighted and the test substance are administered .

### 4. Observation

Observe at least once during the first 30 minutes, periodically during the first 24 hours (with special attention during the first 4 hours), and daily thereafter, for a total of up to 14 days.

### Signs of Toxicity should be Recorded

Increased motor activity, anesthesia, tremors, arching and rolling, clonic convulsions, ptosis, tonic extension, lacrimation, Straub reaction, exophthalmos, pilo-erection, salivation, muscle spasm, opisthotonus, writhing, hyperesthesia, loss of righting reflex, depression, ataxia, stimulation, sedation, blanching, etc.

### 5. Pathology

Necropsies must be performed within 16 h after death. All gross pathological changes should be recorded for each animal. Microscopic examination of organs showing evidence of gross pathology in animals surviving 24 or more hours may also be considered.

### 6. Test Report

Must include the following information as appropriate:

- Test substance-physical nature, purity, physicochemical properties and identification data
- Vehicle - justification for choice of vehicle, if other than water.
- Test animals: -species/strain used microbiological status of the animals, number, age and sex of the animals, source, housing condition, diet etc.
- Test conditions
- Details of test substance formulation including details of the physical form of the material administered.
- Details of the administration of the test substance including dosing volumes and time of dosing.

- Details of food and water quality (including diet type/source, water source).
- The rationale for the selection of the starting dose.

**Results**

- Tabulation of response data and dose level for each animal (i.e. animals showing signs of toxicity including mortality; nature, severity, and duration of effects);
- Tabulation of body weight and body weight changes.
- Individual weights of animals at the day of dosing, in weekly intervals thereafter, and at the time of death or sacrifice.
- Date and time of death if prior to scheduled sacrifice.
- Time course of onset of signs of toxicity, and whether these were reversible for each animal.
- Necropsy findings and pathological findings for each animal, if available;
- Discussion and interpretation of results
- Conclusions

## OECD Test Guidelines

The Organization for Economic Cooperation and Development (OECD) has given five Test Guidelines for acute systemic toxicity testing-

- Acute dermal Toxicity (OECD TG 402)
- Acute Inhalation Toxicity (OECD TG 403)
- Fixed Dose Procedure (OECD TG 420)
- Acute toxic Class method (OECD TG 423)
- Up -and-Down Procedure (OECD TG 425)

## Acute Toxic Class Method

Substance is administered orally to a group of experimental animals at one of the defined doses. The substance is then tested using a stepwise procedure each step using three animals of single sex .Absence or presence of compound-related mortality of the animals, dosed at one step, will determine the next step – i.e.

- No further testing is needed
- Dosing of three additional animals, with the same dose
- Dosing of three additional animals at the next higher or the next lower dose level

### Fixed Dose Procedure

Groups of animals of a single sex are dosed in a stepwise procedure using the fixed doses of 5, 50, 300 and 2000 mg/kg. The initial dose level is selected on the basis of a sighting study as the dose expected to produce some signs of toxicity, without causing severe toxic effects or mortality. Further groups of animals may be dosed at higher or lower fixed doses, depending on the presence or absence of signs of toxicity or mortality. This procedure continues until the dose causing evident toxicity or no more than one death is identified, or when no effects are seen at the highest dose, or when deaths occur at the lowest dose.

### Acute Dermal Toxicity

Test substance is applied to the skin in graduated doses to several groups of experimental animals one dose being used per group .Observations of effects and deaths are made .Animals that die during the test are investigated, and at the conclusion of the test the surviving animals are also sacrificed and analysis performed. This test is useful in testing analgesic combinations with dermal administration

### Up & Down Method (Staircase Method)

Two mice are injected with a particular dose & observed for a period of 24 hrs for any mortality. The subsequent doses are then increased by a factor 1.5 if the dose was tolerated, or decreased by a factor of 0.7 if it was lethal .The maximum non lethal & the minimum lethal doses are determined . A final LD50 assay is planned using at least 3 or 4 dose levels within the range of maximum non lethal & minimum lethal doses

This method has 2 types of tests-

1. Limit test
2. Main test

### Limit Test

The limit test is used in situation where the experimenter has information gained from knowledge about similar tested compounds. Indicating that the test is likely to be nontoxic .If there are little or no information about compound toxicity the main test should be performed.

### Limit Test at 2000mg/kg

Dose one animal at the test dose if the animal dies conduct the main test to determine the LD50. If the animal survives dose four additional animals so a total of five animals are tested. If three animals die, the limit test is terminated and the main test is performed. If three or more animals survive then the LD50 is greater than 2000 mg/kg.

Main Test: In those situations where there is little or no information about its toxicity, or in which the test material is expected to be toxic, the main test should be performed.

**LD50 (Median Lethal Dose)**

- LD 50 is the amount of a material, given all at once, which causes the death of 50% of a group of test animals.
- LD50 is used as an indicator of acute toxicity
- LD 50 can be found for any route of administration but dermal & oral administration methods are the most common
- LD50 value depends on the route of administration (intravenous< intraperitoneal < subcutaneous< oral)
- It is an index determination of medicine and poison's virulence.
- The lower the LD 50 dose, the more toxic the substance.

# 23

# Evaluation of Reproductive Toxicity potential of Xenobiotics

*Atul Prakash and Amit Shukla*

## Epididymal Total Sperm Count

The epididymal sperm is obtained as described above, incubated at 35°C, which is the optimum temperature of rat epididymal sperm. The epididymal fluid is then diluted to a volume of 5 mL of pre-warmed (32°C) the spermatozoa are counted by hemocytometer using the Improved Neubaur (Deep 1/10mm. LABART, Germany) chamber as described by Pant and Srivastava (2003).

Total viable cell/ml = (Average number of sperm per chamber) x 103 x (Dilution Factor)

## Liveability and abnormal sperm

Per cent live spermatozoa is estimated by differential staining technique using Eosin-Nigrosin stain (NE) (Campbell *et al.*, 1953). These slides are also used for estimating the per cent abnormal sperm morphologically on the basis of observable abnormalities of head, neck, mid-piece and trail region of the spermatozoa.

The percent live spermatozoa are determined adopting the differential staining technique using Eosin-Nigrosin stain.

- A drop of epididymal semen is taken on a clean, grease free pre warmed glass slide.
- 4-5 drops of Eosin-Nigrosine stain (1% Eosin and 5% Nigrosine in 3% sodium citrate dehydrate solution) is placed near the semen drop.
- Epididymal semen and stain are mixed gently using a blunt fine glass rod.
- After 30 seconds to 1 minute a thin smear is made on a clean, grease free glass slide.
- The smear is examined under oil immersion objective.

A total of 200 spermatozoa are counted in a slide. The stained and partially stained spermatozoa are considered as dead.

## Sperm Morphology

Two hundred spermatozoa (heads only or intact sperm) per animal are evaluated for head and/or flagellar defects by microscopy (100, total magnification). Classifications of individual spermatozoa are: a) normal, b) normally shaped head separated from flagellum, c) misshapen head separated from flagellum, d) misshapen head with normal flagellum, e) misshapen head with abnormal flagellum, f) degenerative flagellar defect(s) with a normal head, and g) other flagellar defects(s) with a normal head. .

## Sperm Motility Assay

Sperm motility is recorded as percentage of progressively motile spermatozoa after the dilution of semen. The diluent (Buffered 2.9% sodium citrate solution) kept at 37°C is added to sperm suspension until desired dilution is obtained sperm motility is assessed by the method described by Zemjanis (1970). A drop of this diluted semen is kept on a clean, grease free, pre-warmed glass slide cover slip applied and examined under high power magnification (40X) of a microscope within 2-4 min. of their isolation from cauda Epididymis.

## Sperm Suspension for Biochemical Assays

### γ -glutamyltranspeptidase (γ GT)

Gamma-glutamyl transpeptidase (GGT: EC 2.3.2.2). γGT as a marker of Sertoli cell function in testes. Gamma-glutamyl transpeptidase is assayed in the cauda epididymal sperm using L- γ-glutamyl-3-carboxyl-4-nitro anilide and glycylglycine as substrates (szasz, 1974). Gamma-glutamyltranspeptidase catalyses the reaction between L- gammaglutamyl-3-carboxyl-4-nitroanilide and glycylglycine to produce p-nitroanilide (szasz, 1974). Sperm suspension is centrifuged (10,000g for 30 min.at 4 oC), 100-150ml supernatant are used forenzyme assay .The increase in absorption at 405nm at 37oC is followed in a zero order reaction and is directly proportional to the enzyme activity.

## Lactate Dehydrogenase-x (LDH-X)

Lactate dehydrogenase (LDH: EC 1.1.1.27). LDH-X constitutes more than 80% of the total lactate dehydrogenase activity in mature spermatozoa. Lactate dehydrogenase (LDH) catalyzes the NAD dependent interconversion of lactate and pyruvate, thus providing an important source of energy for cell. LDH-X is assayed according to the method of Gold berg and Hawtrey (1987) using α-ketovalerate as substrate. Sperm suspension is centrifuged (10,000g for 30 min.at 4 oC), 100-150ml supernatant are used forenzyme assay.The decrease in absorption at 340nm at 25oC directly proportional to the enzyme. Where one unit of SDH activity is defined as 1μmol of substrate transformed/ minute/ mg protein at pH 7.6 at 25°C.

### Sorbitol Dehydrogenase (SDH)

Sperm suspension is centrifuged (10,000g for 30 min. at 4 °C), 100-150ml supernatant is used for these enzymes estimation (Bergmeyer, 1974). The SDH assay is based on inter- conversion of Fructose and Sorbitol. During the reduction of fructose, an equimolar amount of NADH is oxidized to NAD. The oxidation of NADH results in a decrease in absorbance at 340 nm. The rate of decrease of absorbance at 340 nm is directly proportional SDH activity in the sample, where one unit of SDH activity is defined as 1μmol of substrate transformed/ minute/mg protein at pH 7.6 at 25°C.

### Testes Homogenate for Biochemical Assays

### 3β-hydroxysteroid Dehydrogenase (testicular androgenic enzyme) activity:

3ß-hydroxysteroid dehydrogenase (3ß-HSD: EC 1.1.1.145) enzyme catalyze an essential step in the formation of all classes of active steroid hormones. 3ß-HSD catalyzes the conversion of 3-hydroxy-5-ene steroids (dehydroepiandrosterone, pregnenolone) to 3-oxo-4-ene steroids (androstenedione, progesterone) on a single, dimeric protein containing both enzyme activities. The testicular 3β-HSD activity is measured according to the method of (Talalay *et al.*, 1962). The testicular tissue of each animal is homogenized in 15% spectroscopic grade glycerol containing 5 mmol potassium phosphate and 1 mmol EDTA at a tissue concentration of 100 mg/ml. The homogenizing mixture is centrifuged at 10,000 × g for 30 min at 4°C. The supernatant (1 ml) is mixed with 1 ml of 100 μmol sodium pyrophosphate buffer (pH 8.9) and 30 μg of dehydroepiendrosterone (Sigma) in 40 μl of ethanol and 960 μl of 25% BSA (Sigma), making the incubation mixture a total of 3 ml. The enzyme activity is measured after addition of 0.5 μmol of NAD (Sigma) to the tissue supernatant mixture in a spectrophotometer cuvette at 340 nm against a blank (without NAD). One unit of enzyme activity is the amount causing a change in absorbence of 0.001/min at 340 nm.

### 17β-hydroxysteroid dehydrogenase (testicular androgenic enzyme) activity

17 β -hydroxysteroid dehydrogenase (17 β-HSD: EC 1.1.1.62). 17 β-HSD catalyzes the reduction of the low activity estrogen, estrone, into the potent estrogen, estradiol. The activity of testicular 17β-HSD is measured

biochemically using the same supernatant prepared for the assay of 3β-HSD (above) .The testicular 17β-HSD activity is measured according to the method of (Jarabak *et al.*, 1962). The supernatant (1 ml) is mixed with 1 ml of 440 μmol sodium pyrophosphate buffer (pH 10.2), 40 μl of ethanol containing 0.3 μmol of testosterone and 960 μl of 25% BSA making the incubation mixture a total of 3 ml. The enzyme activity is measured after addition of 1.1 μmol NAD to the tissue supernatant mixture in a spectrophotometer cuvette at 340 nm against a blank (without NAD). One unit of enzyme activity is equivalent to a change in absorbance of 0.001/min at 340 nm.

## Testosterone Level

Testosterone concentration in testes homogenate samples is determined by using RIA kits supplied by Immunotech, France. The unknown samples and the standard sample are incubated with labeled testosterone in antibody-coated tubes. After incubation, the contents of the tubes are aspirated and the bound radioactivity is determined in a gamma counter (Packard, USA). A standard curve is prepared with 6 standards. Testosterone concentration in unknown samples is obtained from the curve by interpolation.

# Index

**T**